System Designs into Silicon

System Designs into Silicon

Jan Johansson
*Ericsson Radio Systems AB,
Stockholm, Sweden*

and

John Forskitt
*GEC Plessey Semiconductors Ltd,
Swindon, UK*

Institute of Physics Publishing
Bristol and Philadelphia

British Library Cataloguing in Publication Data

A catalogue record for this book is available from the British Library.

ISBN 0-7503-0114-7

Library of Congress Cataloging-in-Publication Data are available.

Published by IOP Publishing Ltd, a company wholly owned by the Institute of Physics, London.
Techno House, Redcliffe Way, Bristol BS1 6NX, England.

US Editorial Office: The Public Ledger Building, Suite 1035, Independence Square, Philadelphia, PA 19106 USA.

Printed in Great Britain by Page Bros (Norwich) Ltd, Norwich NR6 6SA

System Designs into Silicon

This book, **System Designs into Silicon** describes the stages and the management decisions to be made when carrying out the evaluation and design of a large electronic system when at least parts of the system are to be realised in specially designed silicon chips (integrated circuits).

It does not describe the detailed use of any software design package as it is not possible to cover all design tools available.

If one tool set was selected, the information that could be provided, would either be unhelpful - it is not the the tool set your company used - or out of date as they change very rapidly.

This book instead, deals with principles and methodology of using the design tools.

The book divides into the following sections :-

* Part 1 - Introduction and Background
* Part 2 - Management Information
* Part 3 - Technical Details
* Part 4 - Appendix.

Preface

The total design of a large electronic system from its conception until it is in production, complete with any enclosure or packaging, will cover a range of engineering disciplines.

These will include :-

* The use of Software Tools ;
  for creating design algorithms, architectures, physical layouts ;
* Design Methodology ;
* Knowledge of Semiconductor Devices and Processing ;
* Knowledge of the Process of Integrated Circuits ;
* Encapsulation and Thermal Dissipation Management ;
* Assembly of Integrated Circuits onto various support media ;
* Packaging of all components for User handling.

As the relative cost of development and production of

> Application Specific Integrated Circuits
>
> or **ASICs**, for short,

becomes less, they are finding application in more and more areas of industry, including companies that have never used electronic circuitry before in their products.
Many managers, with no previous experience of such developments are being confronted by a whole new set of problems.
They are also faced by a whole new set of design tools.

This book aims to help in the resolution of these unfamiliar situations by explaining the design options available and highlighting some of the pitfalls that may be encountered.

One of the problems when dealing with electronic design is the language. Unfortunately much of the language used in semiconductors is **'JARGON'**.

All such words will be explained when they are first used and, if the authors feel it necessary, on other appearances to give an unambiguous understanding of a particular description.

All these words plus any others that are deemed to have specialised meanings, will be gathered together in a glossary. The authors apologise for any omissions from their list. Their saturation with this terminology and over-familiarity with certain words, causes them to forget their special meanings.

Due to pressure of work, many busy managers may find it difficult to read all the book. In order to help, the authors have divided the book into the sections, described below. This should allow reference to individual chapters or sections of chapters to obtain information on specific topics without reading everything.

Although knowledge does build from chapter to chapter, where it has been thought necessary, the authors have repeated the data on some topics in several chapters so that cross referring will not be necessary or will be reduced to a minimum.

The more technical aspects of the subject are kept for later chapters and grouped in the section called 'Technical Details'. Although necessary for a complete understanding of the subject, they may be omitted at the first reading.

Naturally information given in any book on subjects that have been developing as rapidly as semiconductors and CAD, will be either out of date by the time it has been published or quickly becoming so.

The authors have tried to write this book so that the information given will remain a valid basis for decision making even when the scale of integration has increased. The level at which system simulation takes place may also have changed radically but the proposed Design Methodology covers the principles of design and the analysis. These should continue to help with decision making into the future.

Acknowledgements

The authors would like to thank their colleages, too numerous to list here, for the contributions that they have made. They have also helped by reading parts of the text and pointing out errors.
Their suggestions on what to include and leave out were also helpful. The authors however must especially thank Ann Beerling for all her efforts on their behalf during what proved to be a long and trying period. Without her help and support the book might not have materialised.

Contents

PART 2

xiv   Contents

FIGURES

xx **Contents**

TABLES

PART 1

Introduction and Background

Chapter 1 Introduction

1.1 The Intention of this Book

The book deals with large electronic systems and the design procedures that are currently available when they are realised in silicon.
Its endeavour is to help the people managing and designing such projects. It will detail the stages involved in the design and help with the making of certain decisions, before and during the design cycle. The decisions they take from the various options available to them, will require the managers and designers to understand the implications of that choice. By understanding the different effects of these options, the managers should become more confident of their decisions. They should become more competent at making these decisions. The correct decisions will lead to the successful conclusion of the design and development of their own company's products, for functionality, timescale and cost.
The book will help engineers rationalise their approach to large system designs by giving them a total view of the design process.

There have been great changes over the last two decades in the semiconductor devices available on which to implement electronic circuits. CAD tools that have been used in their design, have also shown dramatic improvements.
Simulation at various levels is now a well established, integral part of the design process.

The overall performance of Large Electronic Systems will be enhanced by further developments in several areas, both design software and improved technologies.

As it has become possible to include more and more of the functions in one design in a single integration, so it has become possible to think of forming :-

A System on a Chip

This will include not only the digital parts of the system, for which a high degree of integration has been available for several years but also for analogue parts of an ever increasing complexity.

In addition to the obvious improvement, i.e. the reduction in the physical size of the system, normally there will be some or all of the following benefits :-

* reduction in power consumption ;
* increase in dynamic performance ;
* better overall reliability of the system ;
* security of the design function ;
* a significant reduction in total costs.

This book will describe the various techniques available and investigate the implications of each method on the resources of the development team.

1.2 The Silicon Explosion

1.2.1 Early Offerings

Since the introduction of Integrated Circuits into general usage in the 1960s, many technologies have been used.

Often processes have been developments from previous versions but occasionally some were totally new technologies.

Originally ICs were produced on a Bipolar process which was very simple by today's standards and could yield only very small circuits.

Later MOS technology was mastered and the scale of integration began to increase rapidly. This was followed by CMOS technology which has become the dominant technology because of its particular advantages.

Gordon Moore of Intel Corporation predicted that the scale of integration would grow rapidly, the components on a chip would double every year.
This is known as Moore's Law.
The graph, Figure 1-1 compares actual growth against the prediction.
Subsequent events suggest that the law may have been pessimistic. The rate of growth, at times, appears to have been even higher than that originally predicted. As they grew, Integrated circuits have been categorised according to a relative scale which is given below.

SSI	-	Small Scale Integration ;
MSI	-	Medium Scale Integration ;
LSI	-	Large Scale Integration ;
VLSI	-	Very Large Scale Integration ;
ULSI	-	Ultra Large Scale Integration.

Figure 1–1:   Moore's Law

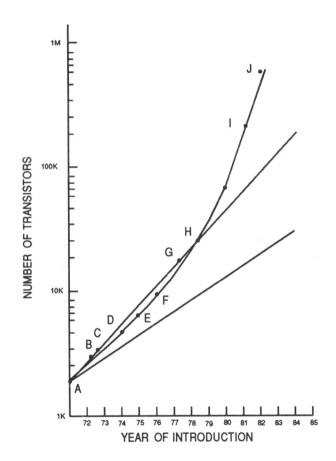

The actual boundaries between the categories are ill-defined but originally were taken as :-

    10   100   1000   10000   gates

In current terminology these may be low.

Today, the predominant technology particularly for new digital systems, is   **CMOS.**

Others technologies available may be used, especially where the systems require higher dynamic performance or special analogue functions, these could include :-

* Bipolar technology ;
* BiCMOS technology ;
* Gallium Arsenide technology.

1.2.2   Measuring the Growth

In the last 10 years, the offerings from semiconductor manufacturers have increased by several orders of magnitude in the product ranges to be discussed in this book.

The scale of integration has usually been measured in terms of

logic gate equivalence

This was because the techniques used were developed mainly in the digital area where the techniques to be described were most readily applicable.

The standard of measurement was normally taken as the GATE.

The usual definition of 'One Gate' is :-

    the number of transistors required to form

    a **Two Input NAND Gate.**

In the case of CMOS technology this is 4 transistors.

1.2.3   Currently Available

The scale of standard products, measured by the terms used in the previous section, is now of the order of millions of gates.

The most visible products with a continuously increasing scale of integration are   MEMORIES.

Larger capacity products are announced regularly, normally with the new devices a factor 4 times greater than the previous ones.

Many other products of this scale of integration are more specific to a particular section of the electronics industry. There will included among those devices that do have universal application a number of general purpose microprocessors.

At present there seems no reduction in the rate of the development in the technologies that allows system designers to include more components on a chip.

1.3   The Unexpected Problems of VLSI

When the semiconductor manufacturers had the capability of VLSI, they encouraged their customers to consider the possibility of using a **System-on-a-Chip** approach to their products.

However there were unexpected financial barriers to this.

As a result of research and development, semiconductor manufacturers were able to produce VLSI silicon. Naturally they wished to exploit this capability. Figure 1–2 shows a possible set of circumstances that could arise because of the development.

Consider the following scenario :-

1. Increasing the Scale of Integration implies an increase in complexity of the product ;

2. Increasing the complexity of a product, normally implies an increase in the specialisation of the function it performs ;
    The product because of these special new features, can be used by a decreasing number of designers. It will no longer have as general an application.

3. Increasing a product's specialisation implies reducing the sector of the total market it can address ;

4. Reducing the market may increase the cost per unit function.
    Chip development costs must be recovered against sales.
    When they are very high, each chip carries a very large overhead unless that cost can be spread across very high volume sales.

Figure 1–2:   VLSI the Problem

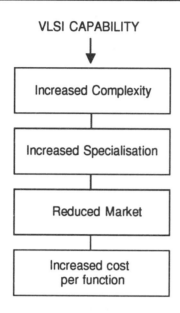

The overall development from design to production of a state-of-the-art chip could take several 10s of man-years of design time and cost several $ millions.
This restricts the potential markets to those standard products that have a multiplicity of applications and hence a wide marketplace.

Such products include :-

large memories and standard microprocessors.

There are a relatively small number of markets where the production volumes are sufficient to justify a full custom design.
These are special situations outside the scope of this book.

1.4   Programmable Silicon

A much larger market exists for customer-specific VLSI where production requirements are modest by semiconductor standards varying from a few hundreds to a few hundreds of thousands.

This potential was the driving force behind the emergence of a range of customer programmable products that were able to meet the financial criteria of both Silicon Vendor and customer.
As already discussed, these products are currently referred to as :-

Application Specific Integrated Circuits - ASICs.

The first offerings were Bipolar and were originally called -

Gate Array or Component Array

There were many other names used by individual companies.

Other products offered included Programmable Logic Arrays and Programmable Memories.

Semiconductor Companies offered ranges of Semicustom products on which the System Houses were given the opportunity to design their own circuits or at least to specify circuits, some very specialised, for their own particular application.

The very high costs of full custom design were reduced by restricting the degrees of freedom allowable.

The techniques available are discussed in detail in a later chapter. See 'Technology and Techniques'.

The first Semicustom Silicon Products had a capability, in bipolar technology, of approximately

200 gates.

Semiconductor companies are now offering CMOS products with a capability in excess of

500,000 gates.
These figures are expected reach 1 million gates very shortly.
See Figure 1–3.
In the case of Gate Arrays, the gate counts quoted are for the core of the array which is fixed for any particular design. In addition to the core count, there are a fixed number of components available for peripheral functions. Each group will be associated with a bonding pad, see the chapter 'Technology and Techniques'.
There may be several hundred pad positions and several thousands of extra transistors and other components associated with them. They are used to create the interfaces between the chip and the outside world.
These components will include bonding pads, gate protection devices and transistors for making a specific pad position either an input, an output, a bi-directional or a power supply.

Figure 1–3:   Growth of ASIC Complexity

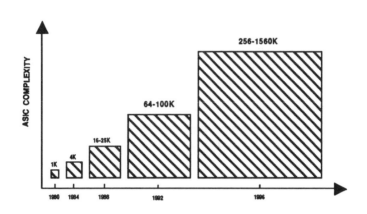

Analogue functions are becoming increasingly available.

Radio frequencies components are needed to operate at a range of frequencies across the entire spectrum, i.e. starting in the short wave band and able to cover the VHF and UHF bands.

This is one of the motivations behind the introduction of Bipolar and CMOS on the same chip - BiCMOS technology.

The other Semicustom offerings, called variously standard cell or cell-based techniques, are more flexible and their interface functions are added to meet the specific needs of the circuit being designed but there are additional costs and, therefore, charges involved.

In Figure 1–4, the increasing average complexity of standard ICs used as components in the construction of a system is plotted against the number of ASICs that would be needed to replace them. The curve suggests that despite the increasing complexity of the ICs used in system construction, the number of ASICs required to replace them actually reduces.
There appears to be a technology cycle of 2 or 3 years duration.
In this period, the complexity that can be integrated on one ASIC increases to between 2 and 3 times the complexity of the previously most complex function.

The cost per unit area remains approximately constant.

Figure 1–4:   Design Complexity

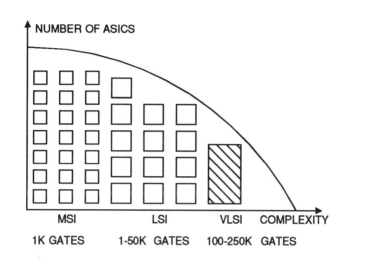

1.4.1   Future Trends

The pace of product development is so great, that most managers find it is impossible to keep abreast of the semiconductor manufacturers latest offering.

As the complexity of the silicon grows, so the support required for the designers working on these increased complexity products must grow.

Unless the design tools being used can be improved significantly, the required level of design support for a 'state-of-the-art' chip will increase dramatically. This is especially true of the later stages where the greater detail will need to be handled mainly by automatic procedures.

1.5   Commercial Pressures

For competitive reasons product development managers must be aware of the various options available to enhance their products. This is true even when there is no immediate intention of using them in the company's product range.

The introduction of advanced semiconductor products by their competitors is likely to have a severe impact on any share a company may have of a particular marketplace.

Any company wishing to keep their present customers and, where possible, to expand their customer base, may be forced to include a higher scale of integration in their products in order to :-

*   reduce costs ;
*   improve system reliability ;
*   improve dynamic performance ;
*   reduce power consumption ;
*   introduce additional product features ;
*   improve their market image.

1.6   The Areas of Impact

There are many inputs to be considered before the correct solution can be found for a particular set of circumstances.

A manager must be aware of the impact each of the inputs may have on the timescales and success of the projects.

```
--------------------------------------------------
Remember, Circumstances change very rapidly.
Available semiconductor products change radically.
The CAD and computing power available for design
may have improved dramatically.
The conclusions reached last time may be outdated
or totally inappropriate for today's product.
--------------------------------------------------
```

BEFORE a decision is made on each project must be reviewed with any locally-important inputs and those listed below :-

*   Can the design produce a correct result in time ?
    No point in starting a design that will be too late !

*   Is the technology to be used mature, proven or new ?
    Are there additional risks being taken ?

*   Is this a 'one off' situation or will further designs ensue ?

*   Is suitably qualified staff available ?

*   Is the available staff trained ?

*   How much does training cost and how long does it take ?

*   Is training readily available or must it be planned in advance ?

*   Is there sufficent computing resource available, mainframe, workstation, memory for all purposes ?

*   How much would a software licence cost ?
    Are there enough software licences to cover the work ?

*   Will the development impact other work programmes ?

1.7   System Development - Background

1.7.1   Breadboard Systems

Historically, Systems have been developed using '**breadboards**'.
i.e. a system created in hardware on a printed circuit board, PCB.
Or several boards in the case of large systems.
In many system developments, this may no longer be a valid alternative on the grounds of complexity or dynamic performance.

The components used in the construction of PCB systems have varied according to when the development took place. They have also depended on the pioneering nature of the company involved. Some adventurous companies have used the latest technology whereas others have adopted a more conservative approach and used only 'tried and trusted' methods and components.

The first breadboards used discrete components, i.e.   single transistors, capacitors and inductors.
These components changed in line with the developments in semiconductor process technologies. ICs including PLDs, of various levels of complexity were used as they became available, moving successively through :-

*   SSI ; MSI and LSI.

1.7.2   A Typical Development

In a product development programme in most companies, before the advent of ASICs, the engineering involved would be covered by a number of quite separate departments.

Figure 1–5:   Company Tree

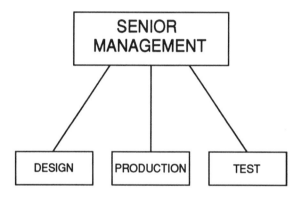

The most common format involved the departments shown in the diagram, communication often occurred along the lines indicated, i.e. only up and down but not across.

Work carried out in all departments was critical to the overall success of the product development but not always recognised as such.

All three are NOT necessarily involved in the early stages of the product development and may have missed important decisions.

Each of these departments normally had a separate development budget and different terms of reference.

The control of the total development budget and the monitoring of the company's overall interest in the particular product's progress took place at a much higher level of management.

In many companies, little contact occurred between the departments except at formal meetings at the level where their management trees converged - e.g. with the Managing Director or President.

1.7.3   Development - Phase 1

The development sequence would start in the Design department.

A paper design would be carried out with as much theoretical calculation as was feasible at the time but based a great deal on practical experience and intuition.

The theoretical results were confirmed using a breadboard.
The interconnections could be wires or printed circuit.

The design might take several iterations before the required functional and dynamic specification were achieved.

The breadboard would be changed by adding components, changing incorrect connections and evolving a product that finally met all the specifications.

1.7.4   Development - Phase 2

The second stage would start with the prototype breadboard being handed over by Development to the Production department.

Frequently, the final prototype with its various, unplanned, modifications, would be unsuitable for volume production.

This would result in further, time consuming modifications until the final product met the criteria for both

PRODUCTION and PERFORMANCE.

1.7.5   Development - Phase 3

The final stage would be with the Test department.

Often at this late stage in the product development, after several design iterations, the testing department would be invited, for the first time, to pass judgement on the suitability - or otherwise ! - of the product from their point of view.

Naturally, as testing had not been a major development parameter until now, there were normally some difficulties to be overcome. This would require additional development iterations before a final, testable product suitable for production volumes, was created.

Typically this would require 2 or 3 further iterations to achieve the final version of the product.

If the overall design was the result of several breadboards then the number of iterations could be dramatically increased, as the various interface problems were overcome.

Frequently the rework could be as long and sometimes even longer than the original design programme.

1.8   System Development - Current Trends

The development of Electronic Systems has been greatly changed over the past 10 years by the availability from semiconductor manufacturers of products that can be programmed to meet the specific requirements of individual companies.
As a result of the semiconductor manufacturers research and development programmes improving their silicon processing capabilities, this trend continues.
Design software has become much more powerful over the same period and now covers many more areas of the design disciplines.
Successive developments have seen an enormous increase in the scale of integration possible.
Design Methodologies have also changed.
The complexity explosion has made it necessary to verify every step of an implementation by simulation.
Otherwise errors may be easily generated and the actual performance of the overall system may prove unsatisfactory.

Initially the number of input and output pins grew approximately in step with the scale of integration.

Packages sizes and pin-counts matched these developments.

The advent of Very Large Scale Integration -   **VLSI**
and the improved CAD however, has brought with it the concept of

A System-on-a-Chip.

A System-on-a-Chip MAY result in a dramatic reduction in the number of input and output connections required.

This can result in a different package problem.

As already stated, until this point in silicon development, there had been a direct relationship between number of gates and pin connections, suddenly there was an imbalance.

Where before the number of pins was often a limitation, now the area of the chip was the determining factor.

The size of the cavity in a package with a small number of pins may be too SMALL to accomodate the relatively large piece of silicon that the chip or die has become.

1.9   Computer Aided Design and Engineering

With the VLSI came extensive suites of -

Computer Aided Design	-	**CAD.**
Computer Aided Engineering	-	**CAE.**

It must be emphasised that this is   Computer **AIDED** Design.

It is NOT Computer Design.

At the present time, if used properly, the CAD can reduce :-

* the physical drudgery involved in the system design ;
* the time overall taken for a design ;
* the number of errors made.

However

the CAD does NOT design the system.

That will still require the designer.

The performance of electronic systems, at various levels of description from 'algorithmic', i.e. System or Behavioural, to 'netlist' or 'structure', i.e. gate level, can be accurately predicted by the simulators that have emerged as part of this CAD explosion.

When necessary, synthesis tools can speed up the design procedure and reduce the drudgery by resolving from one level of description to another for simulation purposes.

These software programs have also changed the methods of the physical design of the chip.

Automatic Synthesis, Floorplanners, Placers and Routers can ensure that much of the procedure has become routine and in many instances, fully automatic.

These tools will only require human assistance to overcome the difficulties of a failure to complete or to make engineering judgements on a compromise between two or more possible solutions.

At the present state-of-the-art, the software will need most help in physical design to achieve high packing densities or to overcome performance difficulties on critical paths.

It can be an advantage for the designer to 'guide' the software by either placing some of the critical logic blocks or deciding the best routes for interconnections.

After the completion of the design stage, other software programs will check the result against the original system design and the technology design rules.

1.10   Design for Testability

The final test of a silicon chip in production can only be carried out from the input and output connections of the package.

Additional access points will NOT be available.

Unless the testing has been planned at the design stage it is very unlikely to produce an acceptable result.

Probe testing using the same test vectors will be carried out on the slice before packaging.

While the slice is still whole, special probes can be used to connect signals and power supplies to the bonding pads that are on the periphery of each chip.

During simulation, the signal vectors that were generated to prove that the system performs to specification, will be assessed for use as the basis for a test program.

This can include any of the self-test methods available such as scan-path, Built-in-Self-Test - BIST, etc.

This will enable the introduction of the concept of

Design for Testability.

1.11 Programmable Design Options

There will be a number of options available to the System designer.

The basic options will fall into two main categories :-

a	software programmable	e.g.	microprocessors, etc.
b	hardware programmable	e.g.	gate arrays, etc.

The details of these options will be described in later chapters.

1.11.1 Software Programmable

An example of software programmable product is a microprocessor.
Programs can be used to produce different responses,
i.e. the function can be changed without affecting the silicon.
The principles of software development and programming are outside the scope of this book.

1.12 Hardware Programmable

Hardware programmable means the standard silicon can be customised to realise a specific system.

These devices will fall into two main categories :-

*   Factory programmable ;
*   Field programmable.

Much of the initial design work will be common to either technique. The final decision on which option to use may be delayed until late in the design process.

1.12.1 General Information

If a permanent change is to be introduced, it should be correct.
The introduction of the **System Design CAD** made this possible.

Two CAD programs were required.

Firstly CAD was needed to help in the design of the system.
Secondly CAD was needed that could check the final version of the design for the correctness of its physical form.

The semiconductor manufacturers introduced libraries of predesigned and characterised electronic functions.
The functionally was verified and the performance of electronic systems accurately predicted on powerful simulators.
This was now possible at several levels :-

* Complete systems represented by Algorithms, with little detail ;
* Netlists with detailed functions made from library models ;
* Physical models using the interaction of diffusion constants.

This meant that complete systems could be accurately modelled prior to any hardware being made. The test vectors used in the simulation could be assessed for their effectiveness in testing the chip. A test program, based on this assessment, should be developed DURING the design phase.

Functional design would be followed by the physical design.
Once again, the design procedure will use powerful CAD programs.
This can now be almost entirely an automatic stage.
Only special requirements of a design, make it necessary for the designer to aid the software.

If the design requires high utilisation of the chip, maximum dynamic performance or both, human input may need to increase before the layout can be completed.

After physical design, a final simulation needs to assess the effect on the dynamic performance of the interconnection pattern.

Before progressing to the next stage in the silicon development, software programs will be used to carry out the following checks :-

> logical and dynamic performance ;
> adherence to the design rules of the particular process.

The designer and the semiconductor company will both have been alerted to any design problems.

1.12.2   Factory Programmable

The next stage in the hardware programmable option involves converting the circuit description data into a format that can be used for writing on a photo-sensitive material.

Two methods are available :-

* Firstly the data can be used for creating conventional masks to complete the processing ;
* Secondly the data can be used to write the patterns directly on the slice or wafer, - E-Beam.

In the first case, if the design is completely satisfactory these masks can be used to process any follow-on orders.
Although quicker initially, the second method provides no permanent copy of the connection pattern. It requires the usage of the Pattern Generator for all subsequent orders.
The use of simulation with only predesigned functions from a library means there is a **VERY HIGH PROBABILITY** the design will be correct and testable first time. If the results have been properly analysed and all warnings heeded.

1.12.3   Field Programmable

After design, the user can take standard devices and in-house give them the required system configuration. According to the type of product, the system configurations introduced may be permanently fixed or they may be changeable.

1.13   Programmable Techniques

Programmable techniques cover the following categories :-

* microprocessors ;
* gate arrays ;
* cell-based or standard cell ;
* PAL - Programmable Array Logic ;
* PLA - Programmable Logic Arrays ;
* FPGA - Field Programmable Gate Arrays ;
* DSP - Digital Signal Products.

1.14   Implications of Various Techniques

Each technique will have different implications for a company.

The final decision on the technique to be used, will depend upon the resources available within the company.

A very important input to the decision will involve the possibility of any future in the company. Is the development a 'once off' need for such a device ? Is it part of a continuous development programme with a regular requirement for many customer-specific products ?

In the former case, the company is unlikely to benefit from undertaking the development themselves.
In the latter case, the company may have much to gain from the learning curve effect. The engineers' experience will be reflected later, in the improvement in the quality of any other ASIC designs completed. There is also likely to be a gain in the efficiency with which it is undertaken.

Every situation will be different even within one company.
It is the intention of this book to help responsible managers arrive at a decision which is the most advantageous option for their own company, at that point in time.

The decision taken may, in fact, change for every system development that is identified.

Changes in the input parameters given below will effect the decision.

*   the point in time ;
    - semiconductor technologies change rapidly ;
*   the development timescale ;
    - will there be time to learn new techniques ;
*   the type of product ;
    - hardware or software programmable ;
*   the quantity and quality of the staff available ;
*   the power consumption of the product ;
*   the cost of the development and production prices ;
*   the quantity and quality of the products required.

1.15   The Design Options

The design options will be :-

* Contract the design to the Silicon Vendor ;
* Contract the design to an independent design house ;
* Carry out the design in-house ;
* Some combination of the above options.

Each will have implications for the System House.

Each option will have a different impact on particular companies.

1.15.1   Design by External Experts

This can be either :-

* the Silicon Vendor ;
* an Independent Design House.

1.15.1.1   Advantages

* The design will be carried out by experienced staff ;
* The design will be carried out on the vendor's computer ;
  - does not impact on other design timescales.

1.15.1.2   Disadvantages

* The design MUST be accurately determined in the specification at the time the contract is signed ;
  This will frequently be very difficult to achieve. The system, often, may not be finalised until well into the design phase. The simulation may show up errors in the initial concept.
* The development charge will be a maximum ;
* The System House will be in the hands of the subcontractor and will not have total control of the design flow ;
* The timescales will be determined by the vendor ;
  priorities will NOT be changed readily.
* No learning curve effect ;
  - your staff will have gained no design experience.

1.15.2   Design In-House

1.15.2.1   Advantages

* The development charge will be minimum ;
* The design will be controlled totally in-house ;
* The development manager will determine priority of resources ;
* Development staff will be gaining valuable experience ;
  - useful in subsequent designs.

1.15.2.2   Disadvantages

* Will use own engineering staff ;
  - will not be available for other projects.
* Staff will require training ;
  - loss of man-hours.
* Design may be carried out by less experienced staff ;
  - increases possibility of inferior designs and more errors.
* Will use valuable computer resource ;
  - will slow down other users.
* Will require software licences ;
  - consumes financial resource.

1.16   Design Productivity

The increase in design complexity, coupled with the decrease in lead times available will require a major increase in design productivity. Will the introduction of new design techniques, either in-house or subcontracted, improve the product development time cycle ?

The first time new techniques are used, will almost always result in a slower design time than achieved in any subsequent developments. It may still show an overall saving on previous methods both in elapsed time and costs.

The saving may NOT be in one particular department but if the new technique can reduced the number of iterations necessary to achieve a correctly designed, testable product, the company will benefit overall.

The timescales and costs of further comparable products will almost certainly show marked improvements.

1.17   Mixed Analogue and Digital Designs

As ASIC technology has moved towards higher levels of integration, system designers have found the need to add analogue functions to the digital circuitry on the same chip.

System partitioning will be made easier if the analogue functions can be integrated with the digital circuitry on one ASIC.

This combined function will normally result in faster, smaller, higher performance, more reliable and cheaper overall systems.

Design tools available for analogue design have always appeared to lag behind those available for the design of digital circuits if measured in terms of transistor complexity handled.

The same perception will also apply to the tools available for layout and test of analogue ICs.

The layout of analogue circuits is much more critically dependent upon the the input of the designer and much less amenable to automatic procedures.

Analogue Test methods and testers also require special techniques.

Chapter 2   Technology and Techniques

2.1   Introduction

This chapter will cover the semiconductor processes that are used to produce silicon that is designed to meet requirements that are either user or application specific, ASICs.

It will also discuss the major techniques used to implement the customers systems on silicon -

Gate Arrays, Standard Cell, Programmable Logic Devices.

These design techniques have the following advantages when compared with full custom :-

using no unproven functions ;

using only characterised functions ;

using only functions verified by previous designs.

This results in benefits such as shorter development times, tighter limits on design parameters and higher confidence in design being 'right first time'.

ASICs are processed by many Semiconductor Manufacturers but may be marketed by even more companies.

The second type of company buys processed wafers from so-called 'Silicon Foundries'.
Silicon Foundries are production lines fabricating standard processes, that sell their slice production against an agreed specification to many customers for individual designs.

The advances that have taken place in the last 10 years in semiconductor technology, particularly in silicon, have resulted in a dramatic change in the size and performance of the ASICs offered.

Customers could now integrate systems of -

> a several hundred thousand gates

instead of

> the few hundred gates

that were originally offered.

In CMOS, the circuits will operate at hundreds of Megahertz instead of the tens of Megahertz originally possible. In Bipolar, the frequency may be measured in Gigahertz.

The complexity of an ASIC is normally measured in gates or gate equivalent, betraying the digital background of this product range.

As mentioned in Chapter 1, one gate is the number of transistors required to form

> a **Two Input NAND Gate.**

In the case of CMOS technology this is 4 transistors.

The first ASIC or semicustom silicon products had a capability, in bipolar technology, of approximately

> **200 gates.**

Semiconductor companies are now offering CMOS products with a capability in excess of

> **500,000 gates.**

In the case of Gate Arrays and PLDs, the gate counts quoted normally relate to the core transistors from which the main system will be constructed.
In addition there will be peripheral components from which the system interfaces will be formed.
These will include the bonding pads that will be used to connect the chip to the pins on the package.

There may be several hundred pad positions and several thousand extra transistors and other components associated with them.
These peripheral components will be used to determine the characteristics of the interfaces that connect the functions designed on the chip to the outside world.
In addition to bonding pads, there will be gate protection devices, transistors and resistors for making specific pad positions a variety of inputs, outputs, bi-directional or power supply connections.

Increasingly there will be components available for making relatively simple analogue functions.
These functions will be required to operate at a range of frequencies, both low and and high, including radio frequencies which will cover the VHF, UHF bands and above.

For standard cell or cell-based techniques, the interface functions including bonding pads will be added as necessary, to meet the specific needs of the circuit being designed.

ASICs are made on standard IC processes with CMOS dominating.

BiCMOS, because it has good HF, interface and analogue capabilities, is establishing a major presence in the ASIC market.
Indeed, the need for better interfaces and analogue functions is one of the main motivations behind the introduction of Bipolar and CMOS on the same chip - BICMOS technology.

Other technologies, used for special areas, are :-

> Bipolar ;
> Gallium Arsenide.

The most important ASIC technique is Gate Arrays.

Other techniques are :-

> Standard Cell or Cell Based Design ;
> Programmable Logic Devices (PLDs).

As the number of useable gates increase, PLDs will achieve a greater market share but are still expected to be at the lower end of the complexity spectrum compared with Gate Arrays.

The decision over which technology and/or technique to use, will be determined by the many inputs to a complex decision matrix.
In many instances where several options could suffice, the ultimate decision will be on price.

The various parameters that may influence the decision will be discussed in later sections.

2.2   Impact of Process Developments

Silicon Vendors will have a continuous - and very costly - programme of process development.

Process developments will have some of the following aims :-

* Increased scale of integration ;
* Improvements in dynamic performance ;
* Changes in operating voltage ;
  both up and down ;
* Reduction in power consumed ;
* Reduction in production costs.

The impact of the first four changes are self-explanatory.

The impact on production costs is much more complex.

Production costs of a process are dependent upon yield.

Yield is the ratio of good / total chips per slice.

Yield is inversely proportion to the number of defects, randomly distributed across the slice or wafer, during processing.

The number of defects - that is flaws in the processed slice - will dependent upon a number of factors, some of the principal ones of which will include :-

* imperfections in the basic slice or wafer ;
* masks defects ;
* misalignment at any of the photo-engraving stages ;
* the presence of dust particles ;
* under or over processing of any stage.

In general, the greater the number of masks and process stages, the lower the yield and the greater the cost per good chip.

Figure 2–1 gives a representation of the variation of production costs with the lifetime of a process.

Initially the costs will be relatively high because the process is new and it needs much attention to make it yield.

As the process procedures become more routine, the yields will improve and the costs fall according to what normally referred to as the Learning Curve Effect.

The mature process will reach a minimum, stable cost.

Figure 2–1:   Variation of Production Costs with Lifetime

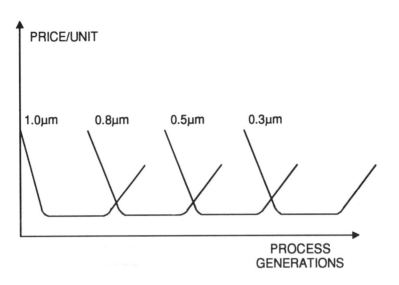

This will continue for a period whilst the process is in full production.

As a process matures it will overtaken by newer developments and will eventually experience a decline in throughput. The costs may well start to increase as it will be used only for old-established products still in demand.

2.3   Comparison of Technologies

In the next sections, the advantages and disadvantages of the various technologies will be given. The comparison tables will be relative to the other technologies being discussed.
All technologies can be supplied to many specifications with regard to minimum feature size, gate complexity, levels of interconnection and power consumption. Individual Silicon Vendors may market several generations of technology.

2.3.1   CMOS

CMOS is the dominant technology for ASICs.

CMOS Technology can be supplied to many specifications.

Gate Arrays have a strong influence in driving the technologies and in this respect are seen to be like memory products.
Dimensions quoted have moved into the sub-micron region.

Other features contribute to the overall performance of a process, important among these features are :-

* metal interconnections technology ;
  including pitch and multi-layers ;
* contact window rules ;
* power consumption.

The turnround time, i.e. the time from the formal agreement to manufacture until the delivery of the first samples, will be much longer for Standard Cell than for Gate Array -

Gate Array	-	1 / 6 weeks
Standard Cell	-	8 / 10 weeks

CMOS ASICs have the following characteristics :-

Most complex products available ;
Relatively low power consumption in most applications ;
Available from many competitive sources - lower prices ;
Moderate operating speeds ;
Moderate analogue functions ;
Moderate low voltage operation ;
Moderate interface capabilities.

2.3.2   Bipolar

Bipolar Technologies can produce high performance ASICs.

Bipolar Technology can be supplied to many specifications.
Sub-micron processes are available.
Supplied by a limited number of semiconductor manufacturers.

Bipolar ASICs have the following characteristics :-

> Moderately complex products available ;
> Relatively high power consumption in most applications ;
> Available only from limited number of sources ;
> High operating speeds ;
> Good analogue functions ;
> Good low power operation in some applications ;
> Good interface capabilities.

2.3.3   BiCMOS

BiCMOS is a marriage of CMOS and Bipolar Technologies.

BiCMOS has the CMOS advantages of complexity and low power consumption with the Bipolar advantages of good interface characteristics, analogue functions and high speed.

BiCMOS can supplied to several specifications with regard to minimum feature size, levels of interconnection and power consumption by a limited number of semiconductor manufacturers.

BiCMOS ASICs have the following characteristics :-

> Complex core features available ;
> Relatively low consumption in the core ;
> High operating speeds ;
> Good analogue functions ;
> Moderate low voltage operation ;
> low packing density ;
> relatively expensive solution ;
> Good interface capabilities.

2.3.4   Gallium Arsenide

Very specialised products that are outside the scope of this book.

2.4   ASIC Techniques

2.4.1   Background

ASIC design techniques have been available for many years.
The first offerings appeared in the late 1960s and early 1970s under a number of proprietary names. They were mainly known by the generic term 'Semi-custom' or Gate Array.

Semi-custom was chosen to distinguish it from 'Full-custom'. Full-custom has implications of :-

* High development costs ;
* Low confidence in success of design ;
    both functionally and performance ;
* Large production volumes.

The semiconductor manufacturers needed to provide their customers and potential customers with access to LSI and VLSI technologies at relatively low development costs and modest production quantities.

The development cost is normally referred to as

NRE - Non-Recurring Engineering.

The basic philosophy demands that

designs are produced correctly the first time.
development costs are reduced to a minimum.

The correct performance of the system function is achieved by the use of good CAD and by restricting the design to using only those functions found in the standard libraries.

The reduction in development costs is achieved in Gate Arrays by having standard processing for all but the metal layers and in PLDs by the user programming standard devices.

2.5   Standard Libraries

For a given technology, the functions used in the realisation of a system design will be constructed from a limited selection of semiconductor cells from the core of the design.
The interface functions at the various bonding pad positions will be another limited selection of semiconductor cells.

In the case of Gate Array and PLDs these cells will occur in fixed patterns on the chips of the base wafers of the product.
In the case of Standard Cell, the semiconductor devices will be of fixed dimensions but their positioning on the chip will be unrestricted.

The basic cells will be used to form a family of basic electronic functions. The functions will have been simulated and verified in silicon before product release. The data will be stored in the libraries supplied by Silicon Vendors to their customers. This data will be used in conjunction with the CAD programs to simulate the performance of a system.

The libraries will contain functions at various levels of abstraction and complexity from very simple to large sub-systems.
These functions are normally proven in silicon.

2.6   Gates

Gates is the most imprecisely used word in ASIC technology.

In Gate Arrays, 'gate' frequently refers to :-

*   basic electronic functions - Nand, Nor, Exor etc ;
*   size of a Gate Array - 2040, 100,000, etc ;
*   group of unconnected transistors ;
*   group of transistors connected to form a function ;
*   method of comparing system complexity ;
    - 5000 Gates equivalent ;
    meaning the system is equivalent to 5000 2-input Nand Gates.

Semiconductor Vendors give their own name conventions to :-

*   groups of unconnected transistors ;
*   groups of transistors connected to form basic logic blocks ;
*   groups of transistors connected as larger, standard blocks ;
*   the special blocks built from the above blocks.

The terminology for the possibilities listed above as used in this book will be defined in the following sections.

Gate will be used to give a measure of complexity.

2.6.1   Array Element

Smallest grouping of transistors in the basic technology.

Not connected to form any function.

In CMOS Technology the most common format is -

> 2 P-type and 2 N-type transistors.

Sometimes the group may be 6 transistors.

2.6.2   Cell

One or more array elements connected to form the regularly-used, basic logic blocks.

These represent the 'core' library of a particular technology.

They have fixed interconnection patterns and will have simulated and hardware characterised performance data.

2.6.3   Macro

Next level of commonly used electronic functions.

Constructed from cells.

Can be 'hard' macros - i.e. fixed geometry including interconnection - producing repeatable dynamic performance.

Can be 'soft' macros - i.e. variable geometry within the constraints of the basic cell shapes. The dynamic performance requires checking after placement and routing to allow for the possible differences caused by the variations in routing.

2.6.4   Modules

Systems or sub-systems built from cells and macros.

May be used once in a design, - e.g. the 'top level' circuit.

May be designed, minimised and optimised on particular parameters, then included many times in the overall design.

Frequently referred to a functional Block.

2.6.5   Parameterised Cells

Some functions lend themselves to automatic design to a size and shape determined by the user.

The design of a '1 bit' version of the function - ROM, RAM, PLA - is stored in the library.
The design specifies the number of bits and words or the boolean expression required.   The software produces the functional and physical design wanted, to the parameters specified.

This is a more efficient method than having a selection of fixed blocks of each type of the functions.

2.7 Programmable Design Options

There are a number of options available to the System designer.

All options depend upon the use of powerful CAD to validate the design against the predesigned, characterised standard libraries.

There is a high probability of the design being correct.

The basic options fall into two main categories :-

        a) software programmable ;   e.g. microprocessors, etc.
        b) hardware programmable ;   e.g. gate arrays, etc.

The details of these options will be described in later sections.

2.7.1 Software Programmable

Software development are outside the scope of this book.

The software development will have been used to program a microprocessor to perform the system functions required.

2.7.2 Hardware Programmable

Hardware programmable fall into two main categories :-

* Factory programmable ;
* Field programmable.

2.7.3 Factory Programmable

These cells receive the special processing necessary to meet the requirements for an individual customer's order before being shipped by the semiconductor manufacturer.
They can not be changed by the user.

2.7.4 Field Programmable

Standard devices that can be given System configurations by the customer, using special equipment on the own premises.

According to the type of product, the system configurations introduced may be permanently fixed or they may be changeable.

2.8   Programmable Techniques

Programmable techniques will be dealt with in specific sections :-

* microprocessors - standard parts - software programmable ;
* Gate Arrays - hardware programmable ;
* Standard Cell - hardware programmable ;
* PAL - Programmable Array Logic ;
* PLA - Programmable Logic Arrays ;
* FPGA - Field Programmable Gate Arrays ;
* DSP - Digital Signal Products ;

2.9   Implications of Various Techniques

Each technique implies different skills, system requirements and resources for a successful development.
The one selected will depend upon the company's resources.
Whether the company can gain from the learning curve effect as the design efficiency improves with the second and subsequent designs will be an important input to the decision. A 'once off' design will offer little advantage to any development programme on customer-specific products.

Every situation will be different and it is the intention of this book to help responsible managers arrive at a decision which, is the most efficient option for their company.
The decision may change for each development in the company.

Changes in the following, effect the decision :-

* the point in time, - semiconductor technologies change rapidly ;
* the development timescale, - is there time to learn these new design techniques ;
* the type of product, - hardware or software programmable ;
* the quantity and quality of the staff available ;
* the scale of integration ;
* the power consumption of the product ;
* the cost of the development and production prices ;
* the quantity and quality of the products required.

2.10   Gate Arrays

At present Gate Arrays are the dominant technique.

There are three main styles of gate array :-

> Channelled
>
> Channelless or Sea-of-Gates
>
> Structured Array.

The only major difference between the first two styles of Gate Array is the presence in the channelled array of large areas of the silicon that contain no transistors.
These channels are allocated for the purpose of running first layer metal interconnections between the various functions being used, second and any subsequent layers of interconnection are not so restricted and can run over most of the chip.

Figure 2–2:   Channelled Gate Array

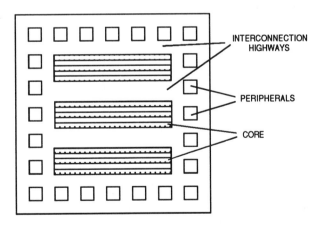

Figure 2–3:   Sea-of-Gates Gate Array

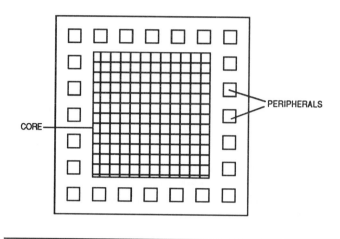

Figure 2–4:   Structured Array

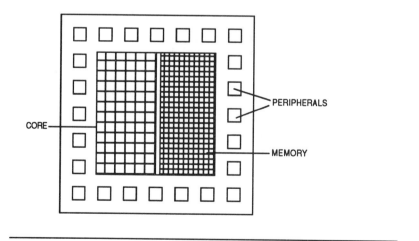

In the case of sea-of gates arrays all interconnections are run over the transistors in the core of the array.

This gives greater flexibility and much better packing density for large regularly laid out functions such as memory. Even so the use of memory is not as area effective as dedicated memory.

This limitation is overcome in some arrays by offering part of the area as Memory and part as normal Gate Array.

This type of array is normally referred to as a 'Structured Array'.

Structured Arrays offer good packing densities in the Memory areas used but some of the Memory area may be wasted.

The Gate array areas as used in the same as similar products.

This difference between the styles does not affect the design procedures, except at the layout stage.

The Gate Array section and Memory being separate preclude the use of 'Imbedded Memory', the memory blocks must be at the side. Communicating between the Memory section and the rest of the system may require the use of multi-bit bus connections.

As the interconnection pattern is frequently the limiting factor in the packing of the chip, more area in the connections between the sections may be lost than is the apparent savings made by the packing in the Memory section.

2.10.1   Chip or Die Size

In the design of a Gate or Structured Array, the final chip size is now heavily dependent upon the area considered necessary to allow the interconnection of the majority of designs likely to be attempted on a particular size of array.

The chip size is fixed for a given technology by :-

*   the process specification ;
    including feature size, metal pitch and contact window capability;

*   the number of gates ;

*   the number of layers of interconnection available ;

*   efficiency of the placing and routing software ;
    including the quality of the programs giving the ability to intervene manually ;

*   the peripheral components.

The number of external connections is also fixed.

A technology update may reduce the minimum feature size and therefore the area for a given number of transistors. As the size of bonding pads will not change in the same ratio, the number of positions that can be accommodated around the periphery will be reduced for a given number of gates in the core.

A match between the system complexity, - number of used gates - and the number of system connections required will determine the size of array needed.

This fixed relationship between the core number of gates and the number of pad connections may result in large areas of the array being unused.

2.10.2   Standard Cell

Standard Cell or Cell-based Design is a design procedure between

Full-Custom and Gate Array.

It has the flexibility of full-custom because there are no restrictions on the number of gates or external interconnections used.

It has the discipline of the gate array because only validated library functions are used in the design.

The NRE for a standard cell will be much higher than a Gate Array because it requires customising at every process step and will have relatively long leadtimes before the delivery of prototypes.

The chip size will be minimum and therefore production prices will be minimum. However there may be little difference in price between Fully utilised Gate Arrays and Standard Cell.

In some Semiconductor Companies the libraries for their gate array and standard cell products are largely common.

The design phase before layout could be the same.
The decision whether to go Gate Array or standard cell need not be made until a late stage in the development.

Standard Cell will only be chosen if :-

the production deliveries are large ;
the design requires special functions not available in the Gate Array library.

Products can start life as Gate Arrays and then for reasons of economy change to a Standard Cell - i.e. to reduce the chip size and hence the cost.

2.11   Programmable Logic Devices

There are many names for the products that constitute this range. Many of them are propriety and appear conflicting.

Each has the facility for customising on the users premises.

They fall into the following categories :-

> Programmable Array Logic ; PAL
> Programmable Logic Array ; PLA
> Field Programmable Gate Array ; FPGA

The differences between the products are in the complexity offered and the manner in which specific system information is entered.
The early products were relatively simple, mainly for use with combinatorial logic and using some form of fusing for writing the pattern on individual chips.
Fuse types write the pattern by a succession of open circuits.
Antifuse types write the pattern by a succession of short circuits.

Later types have used memory built into the chip to accept the system design and to determine the cells used and the interconnection of those cells required.

Some use ROM and some use RAM techniques.
Different ROM technologies are used by different Silicon Vendors and these determine how and how many times the specifying pattern can be written.
Those using EEPROM or RAM to store the system pattern need to rewrite the pattern everytime that the power is removed or to have some form of battery backup.

2.11.1   PAL

Earlist of the PLDs, use fuse or antifuse technology.
PALs consist of a matrix of multi-input AND gates followed by a matrix of multi-input OR gates.
This may be followed by INVERT functions.

The AND functions are programmable.
The OR functions are fixed.

2.11.2   PLA

These PLDs use fuse or antifuse technology.
PLAs consist of a matrix of multi-input AND gates followed by a matrix of multi-input OR gates.
This may be followed by INVERT functions.

The AND functions are programmable.
The OR functions are programmable.

2.11.3   FPGA

These are the latest and largest of the PLDs.
FPGAs consist of an array of array elements, similar to those of other Gate Arrays, with interconnection highways.
The selected functions are placed and connected by the design software and the information stored in a RAM or ROM.
They can be be partly reconfigured - in milliseconds - during the design operation.

They have the flexibility that they may use timeshared sections, the input instructions will determine how and when changes occur relative to the other system data.

Chapter 3   Interface with Silicon Vendor

3.1   Introduction

In order to prevent misunderstanding, the company originating the ASIC design will be called the 'System House'.

A System House that needs to regularly design ASICs and is without an in-house or in-group source of semiconductors, will need to create a good working relationship with an external Silicon Vendor for both prototyping and production.

If a satisfactory interface is to be established, much hard work will be involved in building mutual confidence in each others ability and in creating acceptable working practices.

Once such a working relationship has been satisfactorily established, a switch to another vendor, at whatever level of interface has been selected, will become painful.

If a System House does decide to change Silicon Vendor after starting a design, it will be necessary to overcome a number of difficulties in the difference in the design procedures. The farther a design has progressed the more difficult and more costly it will become to change. Any switch will result in a greater, short term workload and the need for retraining.

This dependency will be greatly reduced in the future as systems are specified in universally accepted Hardware Description Languages such as VHDL - defined in chapter 'Simulation'. The system will be verified at a high level of abstraction and, with certain limitations, could be automatically translated to a range of technology libraries from a number of Silicon Vendors if it is considered necessary. The conversion from the higher level descriptions will require more powerful tools than are at present available.

The feasibility of multiple sources of silicon will be dependent upon the compatibility across both software and hardware.

If when changing to a new Vendor, any major differences in technology and design software are discovered, it will require a confidence building exercise. As the Design House has no track record of success with the new vendor, any doubts about the design will persist until the first, correctly functioning product has been safely delivered. Added to these worries, will be the business pressures that have caused the timescales for the development of new products to get shorter and shorter.

3.2   Choosing a Vendor

There is a great fashion to advocate design software that is independent of the Silicon Vendor. This is in order to avoid a commitment to one company.

Against this argument must be weighed the fact that
Software does not supply Silicon
and the ultimate object of the design is obtaining working Silicon.

For the physical layout and the timing information on a design, it will be necessary to choose the libraries of one Silicon Vendor.
After the experience of a number of designs, frequently a System House will help generate some new entries into the libraries that suit, their own design requirements.

This will encourage a strong commitment by both companies. They will need to work together for considerable periods before proven, debugged libraries become available.

There may be advantages in using the CAD specified by the particular Silicon Vendor because there will be no conflict in the ultimate responsibility for the integration of the libraries into the CAD.

The Silicon Vendors have a great incentive to establish a long term commitment and good working relationship with their customers in the System Houses.

Both will have a vested interest in the accuracy of the modelling producing a very high percentage of circuits that are 'right first time'.

There may be an advantage for a System House to have several ASIC designs in production at one time with the same vendor because their requests for special actions may have more impact.

Should the chosen vendor be large or small ?

This may be a question of flexibility vs stability.

3.2.1   The Small Vendor

Small Vendors will tend to be more :-

* flexible ;
* amenable to small production orders ;
* amenable to pressures from major customers ;
* sensitive to economic variations ;
* vulnerable to market forces ;
* lacking in the resource for developments ;
* influenced by their customers in developments.

3.2.2   Large Vendors

Larger Vendors will tend to be less :-

* flexible ;
* amenable to pressures from customers ;
* influenced by customers timescales ;
* vulnerable to economic variations ;
* likely to curtail development programmes ;
* vulnerable to market forces ;
* likely to listen to customer special requests.

3.2.3   The Ideal Vendor

```
---------------------------------
Remember, the ASIC businesses say
   SERVICE is the KEY to Success
---------------------------------
```

The definition of an ideal vendor would be **flexible, financially stable, innovative and with a good record for service.**

One example would be a progressive, medium-size, merchant semiconductor company within a large, successful organisation.
The lack of any short term viability worries would enable the User and Vendor to build up a close, healthy, long term, business relationship, with good technical contacts at many levels in each company.   Such relationship should give early access to new technologies and techniques - before general release to the market.

The Silicon Vendor should maintain a core set of ASIC Design tools, that will provide all the necessary checking procedures to enhance the flow of designs into silicon but will have a very flexible interface at the front-end of the design route. Libraries will be provided that allow inputs at many levels of extraction, see Figure 3–1.

Figure 3–1:   Design Interfaces

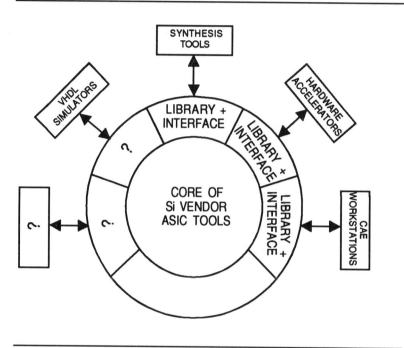

3.3   Design Options

The System House may participate in many of the stages in the design of a silicon chip. In addition the Vendor must be prepared to accept input from their customers that for various reasons have progressed to different stages of the design, see Figure 3–2.

Figure 3–2:   Vendor Interfaces

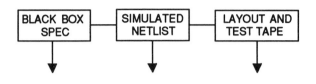

The usual stages are :-

1. System Definition ;

2. System Description ;
          Behavioural ;
          Schematic Capture ;
          Netlist ;

3. System Simulation ;

4. Test Realisation ; Fault Simulation ;

5. Physical Design ; Layout and Routing ;

6. Post Layout Simulation ; including trackloads ;

7. Process Completion is the responsibility of Silicon Vendor.

Stages 1 : 4   may be referred to as ' The Front End '.

Stages 5 : 6   may be referred to as ' The Back End '.

3.3.1   System Definition

The System requirements will be defined by giving :-

          its signal pattern - i.e. input vectors ;

          its expected response - i.e. output signals.

This is frequently referred to as a Black Box Specification.

The detail of the System in between the Inputs and Outputs may not, and indeed, need not be defined by the System House.

This interface may be most realistic if the customer only designs a very limited number of circuits, does not have experienced design staff or does not want to invest in computers and CAD.

It may also be the best solution if the circuit has an extremely critical timescale or technical performance near the limit of any of the limiting parameters of the particular technology.
As discussed there will be an appreciable 'learning curve effect' in the design of ASICs. Second and subsequent designs will normally be carried out much more quickly.
Overcoming the problems of timing on 'critical paths' and high packing density will require experience.

3.3.2   System Description

Some System Houses will prefer to undertake the design of the front end of their system only.
In this situation, the Silicon Vendor must receive details of the system design and simulation results in some mutually acceptable format, based on some form of design review.
It will be necessary to include a suitable test pattern.

The design must have been carried out using the Silicon Vendor's own CAD or an approved third party CAD package with which the vendor has an interface, including approved libraries.

If this design path is to be pursued the System House will have the opportunity of confirming and/or refining the design of the ASIC. The interface with any other components in the overall system will need to be simulated and examined in detail.

The Silicon Vendor will complete the back end design, i.e. layout, post layout simulation, test program generation and processing.
A final simulation including the trackload data from the layout must be completed before the design is signed off for processing and test.

3.3.3   Layout and Test Tape

If a System House has need to have a tight control over critical parameters in many designs per year, it is recommended that the whole design work from specification to completed layout and test tape is carried out inhouse.

This means that the System House must build up inhouse knowledge in all these areas and accept responsibility for all the work carried out.

3.4   Selecting a Vendor

Among the factors to be considered when selecting an ASIC vendor
must be the short and long term objectives of the System House.

3.4.1   Product Development Schedule

The choice will be a proven and an emerging process.

*   What technology will the system performance require ?
    A late choice may give access to a more modern process.
*   When will the process chosen be fully released for production ?
    An early choice may result in an unapproved process.
*   When will the ASIC be needed for System test ?
*   When will the ASIC be needed in production volumes ?
*   Has the CAD to be used been fully proven ?

3.4.2   Important Requirements of Vendors

Technology

*   Performance of the process, - area, power and speed ;
*   Packaging capabilities ;
*   Overall Cost, - NRE and volume price ;
*   Company's Financial performance ;
        Is the company profitable enough to continue in
        existence and with its technology development ?
*   Continuity of Supply ;
        Alternative production facility ;
        Second Source Agreements ;
*   Record of performance as ASIC vendor ;
        Does the company meet its delivery forecasts ?
        Has the company a good record for ASICs that are
        'right first time' and on time ?
        Has the company a good quality control record ?
*   Future Strategies ;
        Are the Vendor's Processes, Industry compatible ?

Design Support

*   Proven Design System (CAD) and Methodologies ;
*   Support of Third Party Tools - Workstations ;
*   Cell libraries offered ;
*   Higher Level Libraries that will support Synthesis ;
*   Education and Training on vendor CAD ;
*   Design support offered ;
        During and Out of Office hours ;
*   CAD Industry Standard Interfaces ;
        e.g. EDIF, VHDL, etc ;
*   Future strategies.
        Are the company's plans for CAD support
        in line the perceived needs ?

3.5   Interface to the Vendor

The System House must develop and prove a strategy that covers
their main operating requirements, for the supply of silicon. This
should meet both their long and short term needs. It must be flexible
enough to cover special situations, if they arise.
It will be necessary to determine which of the interface options given
in the table below will be the normal mode of operation.

The System House must consider the situations that may require
the procedure to be reviewed - how will it cope with :-

*   a shortage of designers ;
*   a shortage of computer resource ;
*   special design features ;
        outside the experience of the present staff ?

	Front End	Back End	Prototype	Production
1.	Vendor	Vendor	Vendor	Vendor
2.	Inhouse	Vendor	Vendor	Vendor
3.	Inhouse	Inhouse	Vendor	Vendor

3.5.1   Route 1. System Definition

Black Box Specification

The System House will give a Black Box specification.
The vendor will design, simulate and physically create the ASIC.

This interface will be optimum if the System House :-

* Only needs a small number of ASIC designs per year ;
* Requires a special ASIC knowledge for a successful design ;
      of the CAD System ;
      the technology ;
* Has a short timescale and a shortage of trained staff ;
* Has a shortage of computer resource.

Disadvantages are :-

* The ASIC must be accurately defined ;
* No design expertise will be gained ;
* The design timescale priorities will not be controlled ;
* Commercially sensitive information may be given away, makes
  the new product vulnerable to competitors products.

3.5.2   Route 2. System Description

Simulated Netlist Specification

Some System Houses may prefer to complete the front end design.
After using approved CAD, they will interface with the Silicon Vendor
at the simulated netlist stage and, optionally, with a floorplan.

The vendor will complete the physical design and process the ASIC.

The advantages are :-

* System House staff will become experienced in ASIC design ;
      Learning Curve effect ;
* The System can be improved as the design evolves ;
      Initial System Definition need not be as rigorous.
* Comparison between pre and post layout simulation
      can be used to influence final design ;
* System House will control priorities during design phase ;
* Will require less computing resource than for Route 3.

Disadvantages are :-

* Will require trained, skilled staff compared with Route 1 ;
* Will require investment in computer hardware ;
* Will Require investment in computer software ;
   Licencing Simulation Design programs ;
* Commercially sensitive information may still be given away ;
   less vulnerable than Route 1.

3.5.3   Route 3. Completed System Design

Layout and Test Tape Specification

System Houses requiring many designs per year and with tight control over critical steps in the design, should complete the whole design.
The interface will be a specification, layout data and test tape.

Advantages of this method are :-

* System House will become expert at ASIC design ;
* System House will control all priorities ;
* The System will improved as the design evolves ;
   Initial System Definition need not be as rigorous ;
* Comparison between pre and post layout simulation will be used to influence final design ;
* The System House will have better use of high level optimisation procedures for investigating other design possibilities ;
* Better confidentiality ;
* Easier to port the design to other vendors ;
* Less vulnerable to loss of commercially sensitive information.

Disadvantages are :-

* Will require trained, skilled staff ;
* Will require investment in computer hardware ;
* Will require investment in computer software ;
   Licencing Simulation and Physical Design programs.

Chapter 4   Packages

4.1   Introduction

After a silicon slice or wafer has completed its processing stages, it has to be tested firstly for its conformance with the process specification and secondly for the functionality of the specific design.

The first series of tests will monitor :-

*   the diffusion parameters ;
*   the interconnection parameters ;
*   the performance of the standard transistor.

This will ensure that any product sold from this particular diffusion batch meets the values in the published process specification, against which all quality control figures will be measured.

The second series of tests will check :-

*   the presence of all required peripheral connections ;
*   the response to the connection of the power supplies ;
*   the performance against the specific test programme.

Those chips that fail the test will be suitably identified, usually with a small ink mark, for rejection later.

Up to this point the chips have been produced in quantity, - there will normally be many of them on each slice or wafer, many wafers per diffusion batch and many batches processed in parallel.

The production costs to this point will be spread over many chips and per chip will be relatively low.

After this point, the chips will be handled individually and the relative cost per chip will now rise.

It is essential for the economics of I.C. manufacture that the chips that go pass this point have a high probability of being 'good'.
Every chip assembled, whether ultimately good or not, will incur the cost of the package parts and the labour for the individual handling.

Yields of 90%+ will be expected.

The chips must be prepared to meet the customers' requirement - ceramic package, plastic package, naked chip, etc.

After visual inspection, a relatively small number of chips may be shipped naked, i.e. unpackaged, for the customer to mount on a suitable supporting medium.
Whereas the vast majority of the chips will be assembled into a package of some description before shipment.

4.2   Single Chip Packaging

Packaging is the process by which semiconductor manufacturers encapsulate a silicon chip in such a way that the user can readily handle it when mounting into electronic systems.
The package will protect the chip from the environment, make available the electrical connections and make it suitable for mounting in the manner chosen by the user.
It will also have an influence on heat management.

After the silicon slice has completed its processing and has been tested for quality and functionality, it is ready for packaging.

The slice will be :-

*   broken down into individual chips ;
    the majority of slices will be sawn with a diamond wheel ;
    they will be sorted on to special carriers ;
    those rejected at test will scrapped at this point ;

*   visually inspected under a microscope ;
    the degree of inspection will depend upon the ultimate specification of the product ;
    military and space products will have longer and more detailed inspection than low cost consumer products.

See Figure 4–1

Figure 4–1:   Assembly Stage 1

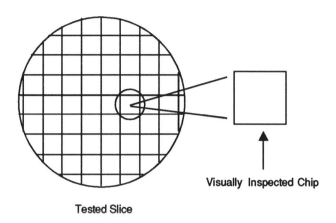

The individually, inspected chips will be either shipped naked or assembled into a package of one of the many types available.

The inspected chips will be :-

*   chip-bonded relative to the leadframe ;
    the leadframe is the electrical bridge between the chip and the medium used for mounting the system ;

*   wire-bonded between bonding pads on the chip and the leadframe with wires ;
    the wires are approximately the thickness of a human hair ;

*   encapsulated, - in a variety of forms ;
    this is sealing the chip and bond wires to protect them ;
    the majority will be injection-moulded into plastic packages ;

*   final tested to ensure the packaged devices still meet the specification ;
    This test may include temperature, environment and power supply variation evaluation.

See Figure 4–2.

Figure 4–2:   Assembly Stage 2

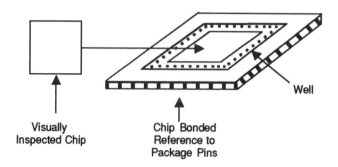

Visually
Inspected Chip

Chip Bonded
Reference to
Package Pins

Well

4.3   Multi-Chip Packaging

```
--------------------------------------------
The abbreviation - SOS - 'Silicon on Silicon'
may be used in this connection but as it is
used to stand for for 'Silicon on Sapphire'
as well, it will not be used in this book.
--------------------------------------------
```

Packaging techniques that allow the mounting of more than one chip
are known variously as :-

MCM - Multi-Chip Module
MCP - Multi-Chip Package

In this procedure, a number of chips of differing designs and differing
processes can be mounted firstly on a substrate of

Silicon   or   Alumina

See Figure 4–3

A multi-layer interconnection pattern of aluminium or copper,
designed to match the needs of the particular module will have been
created on the substrate.
The individual chips will be mounted on the substrate in a similar
manner to single chips in a package.   They will be connected

Figure 4–3:   Multi-Chip Module

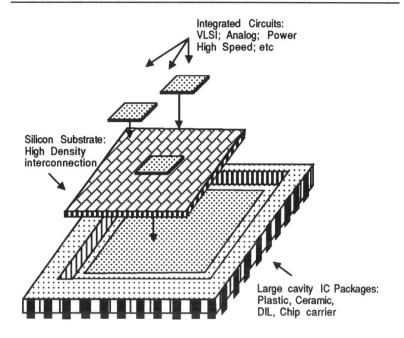

to the substrate pattern either by wire bonding in the manner described earlier, 'tape automated bonding' or 'flip-chip' bonding. A combination of techniques may be used.

After all the chips have been connected the substate, it will be mounted in a package with a suitable well size and, where necessary, wire bonded as before to the external connections of the package.

Obviously this is a procedure that requires high investment in equipment. At present it also takes a relatively long to complete and, hence, is a costly operation but pressure is expected to rapidly erode prices.

This type of assembly has the advantages of :-

*   High Packing Density ;
*   High Quality Interconnections ;
    Transmission Lines if necessary ;
*   Mixed Technology Modules ;
    Analogue and Digital Mixed ;
*   High Quality and High Value Passive Components ;
    Resistors, Capacitors and Inductors, if necessary.

4.4   Choosing a Package

The requirement for well established packages will cause no difficulties but a need for new packages may be out-of-step with the package development programme.
Recently there has been a tendency for progressive ASIC users to be far ahead of the companies developing and making IC packages in defining new requirements.
This may be particularly true of smaller packages.
ASIC users could request unavailable package types.
This may result in a long delay before the package can be obtained and it may also result in a heavy development charge.

Among the factors determining the choice of the package to be used for the production devices will be included :-

*   Chip size ;
*   Number of Input and Output Pins ;
*   System Speed ;
*   Power consumption / thermal parameters ;
*   Cost ;
*   Reliability ;
*   Availability ;
*   Package Mounting Technique ;
*   Fine or Standard Lead Pitch ;
*   Ease of manufacture.

4.4.1   Package Compatibility for Production

Frequently, for economic and timescale reasons, the samples for a new product may not produced in the same package as that used for the production devices. The samples for an ASIC that will ultimately be produced in volume quantities in a plastic package, may be delivered in one of the ceramic options.
It is important, therefore that this possibility has been considered so that the 'bond-out', i.e. the connection order around the package, has been made possible and compatible for the two versions.

4.5   Chip Size

The chip sizes for Gate Arrays will be fixed for a given technology.

The number of pin positions available for external connection will also be fixed.

A match between the system complexity (number of gates) and the number of connections will determine the size of chip.

Cell-Based Design is much more flexible.

The system complexity and interface requirements alone will determine the chip size.

4.6   Number of pins

The ever-increasing scale of integration means that today's designs contain large parts of an electronic system or, in some instances, the whole system.

In general as the complexity and size of chips has increased so has the number of connections required.

Preformed standard packages that are designed to take large chips tend to have large pin counts.

In the case of the chip that contains a whole system, this relationship between gate complexity and the number of connections may no longer be applicable.
There may only be a need for a relatively small number of pins.

There may lead to a mismatch between the size of the 'well' in the package into which the chip will be placed and the number of pins available for connection.

The 'well' is a cavity in a preformed package. A chip can be mounted in the well. It is positioned in such a way that the bonding pads on the chip and the positions on the leadframe can be wire-bonded. That is wire are used to complete the electrical connections between the chip and the external package pins.

The distances between the bonding points on the leadframe and those on the chip have been designed so that, should the bonding wire loops change shape during the later stages of encapsulation or due to temperature variations during operation, any 'sag' in the wire will not cause short-circuits to other exposed parts of the assembly. Standard packages will only takes chip sizes between specified dimensions, a particular sized well can neither accept chips above a critical size nor below a critical size.

In standard packages, the size of chip may be so big that it can only be accommodated in the cavity of package with far more pins than will be needed in a particular case.

The piece parts for such a package may be more expensive than is necessary for the current application.

The alternative would be to pay a tooling charge for a package variant with a larger cavity.

Remember this will take time and, therefore, needs building into any product launch plans.

4.7   Power Dissipation

The general assumption is that all CMOS must be low power.

Some large systems operating at high frequencies using CMOS technologies will consume unexpectly large amounts of power.

An estimation of the likely power the chip will dissipate, will aid the choice of a suitable package.

This is necessary because the performance of Integated Circuits can be unreliable if the junction temperature of the transistors exceeds the value specified by the manufacturer.

The ambient temperature range of the operating system and the temperature rise due to the power dissipated on the chip MUST both be considered in the final choice of package.

4.7.1   Low Power Consumption

In some applications, low power consumption is a necessity.

As there is a direct relationship between Speed and Power, the use of lower power devices is only possible if the implied reduction in dynamic performance is acceptable.

On the same process, low power cells may be physically smaller than standard versions of the same cell.

Therefore low power cells may even generate smaller chips.

The shorter interconnections that result will require less drive and so the overall reduction in performance may be less in this circumstance than first estimated.

4.8   Package Mounting Technique

Packaged silicon chips may be mounted into their operating system using a number of techniques.

Popular techniques include :-

* Insertion into PCB ;
* Surface Mount ;
* Chip Direct.

4.8.1   Insertion In PCB

In this method, the packaged chip will be mounted onto a multi-interconnection layer board in which holes have been drilled. The pins of the typical package make contact with one of the printed layers of interconnect.

Package types that can be used in this way include :-

* Dual-in-line, (DIL) ;
* Pin-Grid-Array, (PGA).

4.8.2   Surface Mount

In this method, as the name implies, the packaged chip will be mounted onto the surface of a supporting medium.
This can be a PCB or a hybrid substrate.

The packages may be :-

*   leadless chip carriers ;
*   leaded chip carriers ;
               a) Ceramic - Cirpak ;
               b) Plastic - Quadflatpack.

LEADLESS packages were developed for hybrid systems.

These packages have similar thermal properties to the hybrid substrates and so track with temperature changes.
If used with PCBs, it will necessary to use a socket or special material in the board construction to compensate for any difference in thermal properties.
Either of which will increase the cost of system manufacture.

LEADED packages typically have leads on all four sides.

When these packages are used, any differences in the thermal properties of the package and the board will be taken up in the flexibility of the package leads.
The pins of the package are formed to make suitable contact with the surface printed layer.
There are two main methods of forming the pins of the package.
These are normally referred to as :-

*   Gull-wing ;
*   J - Form.

Gull-wing pins are formed away from the package
so they lay flat on the surface of the PCB.

Gull-wing packages have low-profile but use more board area.

J - Form pins are folded under the package.

J - Form packages have a higher profile but will use less board area.

Many types of package can be formed for use in this way.

Figure 4–4:   Surface Mount Packages

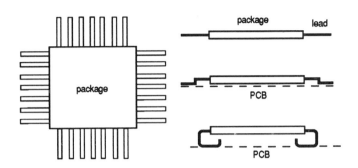

4.8.3   Unpackaged Chips

There are a number of additional techniques used for mounting unpackaged chips directly on a supporting medium.
Some of the more important ones are listed below.

* Flip Chip ;
* Tape Automated Bonding, (TAB) ;
* Advanced technologies ;
  Wafer scale integration.

This is a specialist area and outside the scope of this book.

PART 2

Management Information

Chapter 5   Designer's Profile and Design Productivity

5.1   Introduction

The success of the design of complex electronic systems will be very dependent upon the designers.
They will need to :-

* be of the right calibre ;
* have received suitable training ;
* have access to suitable experience ;
  their own or that of someone else.

This chapter looks at the ideal profile of the engineers that will be undertaking the system designs of the future.

As the amount of effort required at each level in a design varies enormously, it will be necessary to understand where in the design cycle the greatest system improvements and the greatest savings in manpower can be made.

The impact on design productivity in this context will be considered.

The designers and the manager must understand the

capabilities and limitations

of themselves and the design tools being used.

It is necessary to emphasise that it is

Computer **Aided** Design.

The designer will create the design by using the tools sensibly.

5.2   Manpower Utilisation

In the future, most large electronic system designs will in the initial stages, follow a 'Top-Down' approach through the system design stages, shown in Figure 5–1.
The meaning and an interpretation of the stages the design can follow are dealt with in a later chapter called 'Hierarchical Partitioning into Abstraction Levels'.

The levels represent not only the tasks to be undertaken but also the effort that will be required at each stage.
As the design progresses down the pyramid, more detail will be disclosed and more effort required in order to complete that level.

Figure 5–1:   Work Pyramid

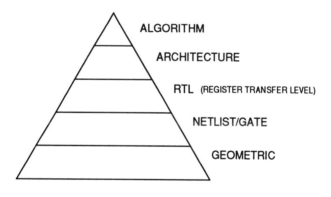

Recent studies suggest, that the effort required is related to :-

> the length of the base of the level
> raised to a power between 2 and 3.

i.e. as the design progresses down the work pyramid, the increase is NOT linear but somewhere between a square and a cube law.

The actual value will be determined by the nature of the system being designed, well-structured systems taking the lower figure.

Moving the analysis and verification of the system to a higher level of abstraction plus using more automatic design tools will change the shape of the diagram.

This change will bring improvements by reducing the amount of manual effort needed at the lower levels of the design and may also improve other aspects of the design.

This will be discussed later in this chapter.

5.3   Profile of a Future Designer

This scenario for a future design shows that there will be a need for greater interaction across the stages of the design.

Figure 5–2:   Designers Profile

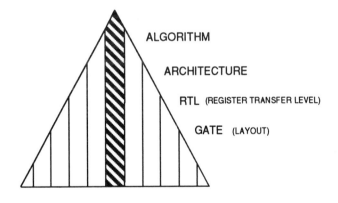

In an ideal situation, future system designers will have a

tall, thin profile.

See Figure 5–2, the designers will have knowledge from

system inception to physical layout of the ASIC.

It will be difficult to fill all positions within a project team with designers of this calibre.
However each team will need at least one of these special people.

The major duties of this designer will be to perform and control the optimisation that takes place over several levels of hierarchy.
This designer will be the communicator between the work being carried out at the various levels in the design.

The designer does not have to be a specialist in all disciplines.
However, knowledge of all areas and experience of working in as many as possible will enhance the overall success rate of the projects undertaken.

The designer, when necessary, will acquire the detailed knowledge from a support group, the membership of which will include specialists covering all steps in the design flow.

The task of the support group will be to acquire, validate and transfer important design information and techniques. These extra skills will be used to help guide the optimisation of the system design over several hierarchical levels.

The task for the support group will be to achieve maximum productivity and to ensure that all designs are 'right first time'.

Software-oriented designers will more readily write programs describing high level system behaviour.

As traditional ASIC design is boring after one or two designs, the 'tall system designer' approach will probably increase the interest in ASIC design by widening the view taken, as well as minimising the need of extra coordination management.

These special designers will assume responsibility for a large part of the project and much of the coordination.

The ideal working environment with the relative disposition of designers and support team is shown in Figure 5–3.

All designers will not all have the same skills and experience.
The Support Group will provide the continuity in the design flow until the necessary experience has been gained.

Figure 5–3:   Designer / Support Group Matrix

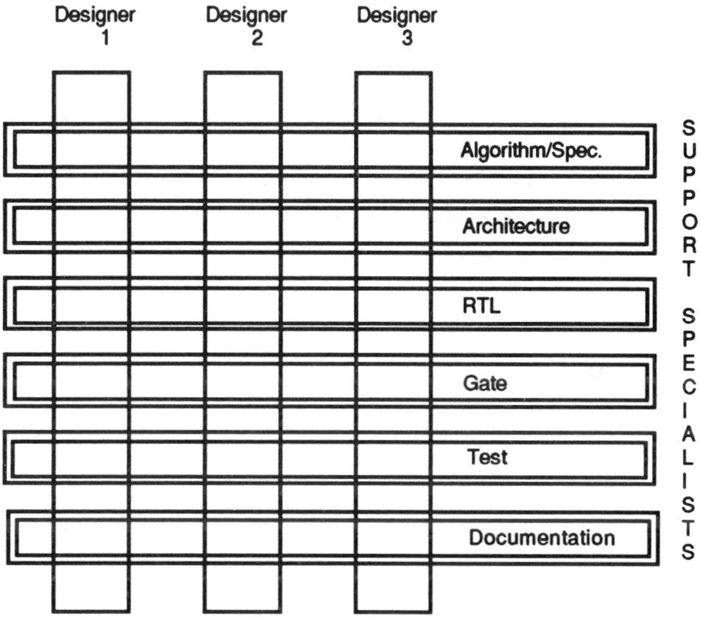

5.3.1   Experienced Designers

In the use of advanced technologies and techniques there will be, as usual, a learning curve effect.
That is, in gaining experience of using the new tools, the newcomer will take longer on the first design than an experienced person.

Experience suggests that the productivity will go up by a factor of 2 or 3 on a designer's second ASIC. Therefore for the efficiency of a design group, it will be necessary to try to keep some designers for several designs.

It will be important to make the design process interesting, creative and challenging enough to achieve the object of keeping designers for a second and subsequent designs.

5.4   Design productivity

5.4.1   Introduction

As the complexity of ASICs increases, the amount of work is bound to increase, therefore the only way to maintain the time relationship of any design to market, or better still, improve this timescale, will be to increase the productivity of the designers.

A typical design flow for a top_down methodology will be :-

*   definition of the algorithm ;
*   verification of the algorithm ;
*   implementation of that algorithm into an architecture ;
*   verification of the architecture ;
*   resolution to the Register Transfer Level ;
*   verification of the system at that level ;
*   resolution to netlist or gate level ;
*   verification of the system at that level ;
    including physical design and resimulation with track loads ;
*   generate Test program.

However the design team and the design task are organised, in any drive for Design Productivity, remember that :-

*   the designer must be given enough computing power ;
*   CPU power is cheaper than engineering time ;
*   in a shared computer mode, the higher the utilisation,
    the more CPU power is used in managing the system.

Traditionally optimisation for :-

*   the dynamic performance ;
*   chip size ;
*   power consumption.

has been attempted at the later stages in the design sequence.

The next section will described how this is no longer acceptable and optimisation MUST be attempted at earlier stages of the design.

5.4.2   Measuring Productivity

The traditional way in which design productivity was measured, was in terms of

gates per man-week.

Although this still has merit for simple designs, it is no longer applicable, or even meaningful, when dealing with very large designs with automatic generation of the system detail and test programmes from the behavioural or architectural descriptions.

Some other standard will be required.

Typical design productivity for today's ASICs using currently available CAD with the present design methodologies

- both Gate Arrays and Standard Cell -

- from system concept to a Layout and Test Tapes

will be in the range :-

500 to 5000 gates per man-week.

The actual value will depend upon the nature of the system, the level of abstraction used for verification and the synthesis tools available for completing the design.

Random logic may take much longer than structured logic but synthesis may reverse this trend in the future.

Any new measure of productivity will need to include elements that relate to both

quality as well as quantity.

Designers that can produce systems that require little or no optimisation will have a higher productivity rating than others who have been much less successful in this task.

As was stated earlier in this chapter, the Learning Curve effect will be an important ingredient in any timescale devised.

The system designers will have to accept more responsibility for the timescales of the whole project and not just the part in which they are directly involved.

5.4.3   Optimisation

The following figures are typical for optimisation on a large, custom design, - 100,000 + gates.

At the bottom level of the work pyramid, where primitive cells or 'polygon pushing' for the final layout are being dealt with, optimisation work may only yield improvements of the order of 10 - 30 % on the chip size, critical path delays and/or power consumption for one man-year of effort.
This contrasts sharply with possible savings, for as little as one man-month effort, of as much as 50 % in the chip size, power consumption and/or dynamic performance if the optimisation has been started at the algorithmic level.

Optimisation effort should be concentrated early in a design.

The nearer the top of the design pyramid i.e. the earlier in the design flow the optimisation takes place, the higher is the potential improvement in both performance and productivity.
It can be shown that between 70-80 % of the performance and, often cost, of an ASIC can be determined during the algorithmic and architecture phases in the design flow.
Whereas traditionally the majority of work involved in an ASIC design was in the later stages as illustrated by the work pyramid,
see Figure 5–1, in the future, the pyramid should be inverted.

As the systems being implemented become far more complex, these figures for system implementation MUST be increased by several orders of magnitude to satisfy the future needs of design productivity.

5.4.4   Conclusions

There will be a number of major factors that can help in improvement of the productivity in the design of large electronic systems.
They include the use of :-

*   Incremental verification of the System Implementation ;
    Behavioural Description with a High-Level Description Language such as VHDL ;
*   Synthesis from Higher Level Descriptions ;
*   Hardware Accelerators for Functional and Fault Simulation ;
*   Automatic Test Program Generators ;
*   Automatic Layout.

Any company using good methodology, effective tools including synthesis, powerful computers and experienced multi-disciplined designers will achieve an appreciable reduction in development time and hence development cost.

Normally in a market driven situation, the timescale overrides other project considerations. This may result in a conflict on budgetted costs as overspent attempts to pullback time lost.

The design productivity will be enhanced by the following rules :-

* Use a proven technology ;
* Use proven CAD ;
  including automatic tools for :-
  synthesis to reduce work at lower levels ;
  test pattern where appropriate ;
  placement and routing ;
* Use proven hardware platforms ;
* Have sufficient computer resource available ;
* Educate and train all the designers in all skills they need BEFORE the project starts ;
* Use (or reuse) experienced ASIC designers ;
* Have support specialists available during the project.

Chapter 6   Design Support

6.1   Introduction

When a System House first makes the decision to use some form of ASIC in its product range, it will be necessary to assess the types of devices available and the current status of the market.

The System House will find it necessary to establish contact with a number of Silicon Vendors to discuss characteristics and performance of their present offering and their planned future product ranges.

It will be necessary to investigate what design procedures they currently favour and which other design software they support.
It will be necessary to know on which hardware platforms the programs are run and the preferred mode for any data transfer.

The prudent System House will also contact the independent CAE suppliers to assess the many options available from them.

The course of action followed by a particular user will depend on how the following questions are answered :-

* the number of ASIC designs needed ;
* the timescale over which the designs will last ;
* the availability of suitable designers ;
* the availability of suitable hardware ;
* the company's financial situation.

If the decision should come out in favour of the System House undertaking the designs in-house, it will be necessary to establish a core of expertise in some sort of ASIC design support team.

It will be not necessarily in the interest of the System House that the system designers who will ultimately design the ASIC, should establish these contacts with either the Silicon Vendor or the CAE suppliers.

For the following reasons, these contacts may be best carried out by experts.

Firstly, lack of experience might make the choice hazardous.

Secondly, lack of time for a thorough investigation may make the choice far from optimum.
The design procedure chosen must be effective and compatible with existing design procedures, both for hardware and software.

Thirdly, the design staff have other duties to perform which may give the System House a better use of their valuable time.

6.2   Assessing the Options

In a large Company in the interest of efficiency and uniformity it is normally advantageous to use a common design system and a common Silicon Vendor.

It will be uneconomic for ALL designers from ALL projects to learn to drive ALL design systems themselves.

The assessment may best be carried out by a support group.

The design support can be supplied by :-

*   the Silicon Vendor ;
*   a CAE vendor ;
*   by a special group internal to the System House.

The option chosen will depend upon the number of designs required and what ASIC design strategy has been chosen.

6.2.1   Design Support from the Silicon Vendor

This strategy will require close contacts between the System House and the Silicon Vendor in two areas.
The support on computing topics should cover :-

*   Recommendation on the computers to use and their setup ;
*   Training on the ASIC design software ;
*   Communication with third party workstation design tools ;
*   Estimation of required resources for completion of the ASIC ;
*   Responsibility for and response to bugs found in the ASIC libraries and/or the design tools ;

Other aspects of ASIC design :-

* Technology libraries, Advice on technology selection ;
* ASIC design rules ;
* Supply of design manuals and design reviews ;
* Selection of package ;
* Reliability information ;
* Interfaces to test machines ;
* Success record of 'right first time' designs.

6.2.2   The CAE Vendors

The check list will be similar except for the Technology aspects. There will, however, be a requirement to stress :-

* The interface with the Silicon Vendor chosen ;
* The responsibility for bug fixing.

6.2.3   Inhouse ASIC Design Support Group

This group must work closely with the Silicon Vendor, CAE Vendor and their own system designers.
The System House management must define :-

* What technologies need to be supported ;
* Which vendors software and on what hardware ;
* The terms of reference for the interface to support group.

The support group should do at least one complex design per year to keep in touch with the real problems in an ASIC design.
The support group should also cover :-

* ASIC design information and help desk ;
* Updates and maintenance on all ASIC design tools ;
* Education on the ASIC design software ;
* Teaching design methodology ;
* ASIC vendor survey ;
* ASIC design tool vendor survey ;
* Develop new ASIC design methodologies.

6.2.4   Design Support Group

An inhouse support group that can cover the roles listed above and can support 10 to 15 complex ASIC designs per year typically would include 10 people.
They should be a mixture of :-

* System Design Specialists ;
* ASIC Design Specialists ;
* Software Specialists ;
* Hardware System Specialists.

6.3   Total ASIC Support

The Design Support Group must assist in :-

* Choosing the most suitable ASIC type ;
* Choosing the Silicon Vendor ;
* Choosing the Design Software ;
* Solving Design Problems ;
* Developing a suitable Design Methodology ;
* Improving the Design Efficiency.

6.4   Total Design Support

During the course of any large system designs, it is inevitable that there will be software 'crashes'.
The design process will stop and output various error messages. These may relate to breaches of correct procedure that fall into a number of categories, namely those related to :-

* the computer's basic operating system ;
  including settings of the computer's operating parameters ;
* the software design tools ;
* the design data.

The typical ASIC designer will be unprepared to deal with this. Inexperience will make difficult to identify the cause.

The Design Support Group must interpret and correct the errors. They must generally be able to overcome any of the difficulties and deficiences that may be encountered during the design sequence.

Not only must they interpret the error messages but they must also be able to provide the solutions.

The support for the ASIC designers, whether supplied by an inhouse group or external experts, must cover the following :-

Training	-	ASIC Design Tools
	-	Design Methodology
Helpdesk	-	Using Design Tools
	-	Design Rules
	-	Problem Solving during Design
Software	-	Installation of Design Tools
	-	Maintenance of Design Tools
Hardware	-	Setup Computers and Network
	-	Make available adequate Disc Storage
	-	Provide Regular Security Backup Facilities
Design Review 0	-	Project Risk Analysis before Starting
	-	Provisional Specification
	-	Chip Plan
Design Review 1	-	Select Design Route and Test Strategy
	-	Select Technology and Package before Starting Design
Design Review 2	-	Verify after Functional Design
Design Review 3	-	Verify after Physical Design
Design Review 4	-	Analyse Completed Project
	-	Feedback on Difficulties
	-	Assess Quality of Support

Chapter 7   Analysis of Project Risks

7.1   Introduction

Before the start of any major design, it is advisable for the management to carry out an analysis of all the risks inherent in that design.

The project manager should try to predict the probability of the project's success in the given timescale.

This analysis should include the impact on the company's product plans of any failure to achieve the agreed target figures.

Those areas of the project where the risk of failure will be at its highest, include :-

* timescales ;
  inability to achieve critical dates ;

* technology ;
  detailed design exposes that one or more parameters will not meet the specification for the project ;

* personnel ;
  designers not available or fail on technical problem ;

* costs.

Some aspects of the project will not be under the control of the System House manager, others will be a direct responsibility.
The difference must be clearly understood.

In those areas not controlled by the System House, it will not only be necessary for the project manager to understand the possible risks but also who does have responsibility for their control.

In the other areas, the project manager should determine totally, all priorities of the project.

The System House will normally have no control over the priorities and changes associated with the technology.

The choice of a suitable Silicon Vendor will be made normally based on previous experience of their delivery and the quality of their products or on their reputation for these services.
Once the choice has been made, delivery dates and silicon performance are no longer under the control of the project manager. There will still be risks, the Silicon Vendor may fail to meet their current promises on the delivery of ASIC samples or their past achievements in terms of its performance.

Other risks will depend upon resources available, relative to the complexity of the project and its overall importance to the company.

Further important areas of risk for the project will be :-

*   motivation of personnel ;
*   administrative support.

The evaluation of the resource allocation should identify the inherent risks before the project starts.

The impact of any failure should be assessed and contingency plans agreed, according to the priority of the project.

7.2   Product Specification

A stable, approved Target Specification
is an absolute necessity before any meaningful review of the risks involved in a project can be undertaken.
This should, at least, define the complete function, in the broadest terms, including the dynamic performance required.

The accuracy of any risk forecasts will be determined by the detail that is available at the time of the review.

Remember that even in CMOS technologies, large systems will dissipate relatively large amounts of power. The thermal properties of the package, the ambient operating temperature and the predicted rise in the junction temperature of the ASIC due to the power dissipated on the chip must all be considered in the specification.

7.3   Motives for the Project

Why design and use an ASIC at this time ?

There are many reasons why a System House should undertake an ASIC design, each will effect the risk analysis in a different way. Some of the possible reasons are given below :-

* Enhancement of the performance of an existing ASIC ;
* Replacement of an existing system module or complete system with a higher scale of integration ;
* Design for a new product.

The category of the product under analysis will modify the answers to the questionaires in later sections of this chapter.
This may change the response to the project review.

Remember if the development is due to a crisis, the effort demanded may distort the entire programme of the design department !

Any launch of a new product that includes the ASIC design under review will be seriously affected by any delay in the ASIC. The launch may miss the critical time for new products in its market.

When the ASIC replaces some other means of manufacturing the product, any delay in the arrival of the ASIC may not be so immediately disruptive to production.

7.4   Motives for the Project Review

Project assessment should take place at many management levels.

The need for these evaluations will differ according to the level in the management tree at which they take place.

The project review may be undertaken to :-

* establish the competence of the project management ;
* guarantee the support for the project ;
* ensure the established design methodology can be followed ;
* identify critical areas and approve the contingency planning ;
* alert senior management to any risks in the project.

7.5   Project Timescales

The development programme for an ASIC should start with the target date for the approval of silicon, 'working to specification'.

The target for each stage should be agreed, based on this date. The critical stages, in reverse order, will determine the overall lead time, necessary to get a product into production.
This works back from the date needed for production deliveries.
The answers will fall into two categories,

> fixed times, determined externally ;
> effort required to complete a particular design stage.

The milestones could be :-

Production Lead Time	Weeks
Analysis of Samples	Man-weeks
Delivery of Samples	Weeks
Design Review	
Re-analysis of Simulation, including Layout Parameters	Man-weeks
Physical Layout	Man-weeks
Design Review	
Analysis of System Simulation	Man-weeks
Entry of System Description	Man-weeks
Design Review	
Definition of System Behaviour	

In the end, it will be necessary to deal in elapsed time.
However the stages where the time is given in 'Man-weeks' may, according to the priority, be allocated additional resource to achieve the required timescales or to redress the situation in the case of failure on any one target date.

Some milestones can not be influenced by higher priority in-house, e.g. the agreed dates for delivering data to the Silicon Vendor.
They are fixed externally.
This analysis will determine the elapsed time needed to meet the targetted delivery of silicon samples and production volumes. The impact of failing to meet this, may impinge on other target dates associated with the product launch.

7.6   Technology

The timescales and the performance of the silicon are the province of the Silicon Vendor.

The Silicon Vendor may accept the contractual responsibility for late delivery and non-performance of the ASIC.
There will nevertheless be an impact on the timescale of the whole project by any delay in achieving correctly working samples.
The responsibility for late delivery and/or non-performance of the ASIC, will be irrelevant if the delay totally disrupts the timescales of the whole project.

If most cases, the project manager should have built into the timescale of the project the possibility of a rerun.
The evaluation of the need for allowing time for a possible rerun for samples will be as the result of answering a questionaire similar to that given below. The answers should assess whether, in the current conditions, this ASIC project is vulnerable or is likely to be successfully completed to the planned timescale.

* Are there any risks in the technology chosen ?
* Will the ASIC be near to the limit of any parameter ;
  Will the packing density be above average ?
  Will the ASIC be expected to operate faster than normal ?
  Will the ASIC be expected to consume power near to the package limit ?
* Will the design data be ready on the agreed date ?
* Will the silicon be 'right first time' ;
* Will the silicon be of acceptable quality ;
* How long and how much will a rework take and cost ;
* Will all the budgetted costs be achieved.

Individual project needs should be included in the questionaire.

7.7   Personnel

The assessment of the suitability of the personnel available for the project will fall into the following categories :-

* quantity and quality ;
* experience and expertise.

It will be necessary for the project manager to obtain answers to the following questions :-

*   Are sufficient design resources allocated for the project ?
*   What success has the project leader had in ASIC design ?
*   Are the designers experienced in digital system circuitry ?
*   Will analogue circuitry design be a problem ?
*   Do the designers have experience in ASIC design ?
    If not is training planned for the beginners?
*   Do the designers have experience in use of simulation ?
*   Are the designer experienced in 'Design for Testability' ?
*   Does appropriate Design Support for this Project Exist ?
*   Are holidays planned for ?
*   Are any of the designers committed to other activities, with comparable priority ?
*   Are any of the designers likely to leave the company ?

The manager will then be in a position to determine whether the balance of the team will be correct for this project.

7.8   Design Resource

The design resource will fall into two categories.
The first category will be available hardware.
The second will be the usage of the hardware.

Firstly answers will be needed for the following type of questions :-

*   Are the CAD tools and hardware platforms appropriate ?
    i.e. do the design tools support the current circuit complexity ?
*   Does the computer have enough local memory and mass storage for the complexity of this design ?
*   Can the hardware platforms manage the circuit complexity without frustrating the designers ?
*   Can the hardware platforms support all aspects of the proposed design methodology ?
*   Are CAD resources like software licences, installations and maintenance allocated for the project ?
*   What test equipment is available and how is it supported ?

Secondly it will be necessary to obtain answers to questions on various aspects of system design :-

*   Can the simulation include the working environment ?
    i.e. can all the interfaces be simulated ?
*   Has the project a difficult timescale ?
*   Will the methodology support the complexity of the design ?
*   How much of the ASIC is working close to the performance limit of the selected technology ? Special aspects may need more manual work than originally planned.
*   How complex is the ASIC, in gate count ?
    10k, 20k, 50k, 100k, 200k, 500k.
    The relationship between gates and workload is NOT linear.
    Complex ASICs need more work, relatively than simpler ones.
*   What is the ratio of random logic to structured logic ?
*   How much of the system can be generated by synthesis ?
    This will influence the manpower needed.
*   How many analogue functions are required for the chip ?
    Analogue functions always make the design work more complicated and take longer to complete.

Chapter 8 Design Reviews

8.1 Introduction

A design review is a control procedure produced :-

* to support the selected design methodology ;
* to ensure that the designs are adequately checked ;
* to identify any areas of risk ;
* to improve the probability of 'right first time'
  when designs are committed to fabrication for prototypes.

Design Reviews should work as a design audit.

They should ensure that the design methodology and design rules have been correctly followed.

They should also close the design cycle loop. If any difficulties are identified in current design practice, discussions including the the designers should propose solutions. That will result in improvements to the design methodology.

Design Reviews should take place :-

* internally at the System House ;
* between the System House and the ASIC Supplier.

8.2 Aspects of Design Reviews

Design Reviews should provide detailed records of ;-

* the design procedures followed ;
* design tools used ;
* versions of software used ;
* the history of developments up to production release.

Design Reviews should be designed to alert the designer in good time, i.e. before too much detail is covered, of special requirements in parts of the design, i.e. non-standard applications.
The ASIC philosophy is based on the assumption that ONLY existing, proven functions should be used.
Normally, Silicon Vendors are unhappy about supplying certain functions and will usually require notification of the presence in any design of these particular configurations.
Included in the list will be :-

*   many analogue functions ;
*   one shot monostables ;
*   long divider chains.

These functions will be included on the list because they will be difficult to test or because they will be less predictable in their performance than the main library items.

The design review will alert the Silicon Vendors to these potential hazards and will allow them to introduce checking procedures.

A design review will typically be a check list of questions.

These questions will have been designed to determine whether the user inputs have been defined correctly and to investigate the extent to which these requirements may conflict with the needs of manufacture and test.

There must be high confidence that the design will satisfy all system requirements.   Any ambiguous aspects of the design must be removed at the initial level of abstraction before moving down the hierarchy to a more detailed level of implementation.
In complex designs, experience has shown, that confidence will be established using hierarchical system verification using high level functions, therefore the use of design reviews should be introduced at the highest possible level in the system specification.

Each design review should always cover timescales.

The system designers and support staff experts should both agree to the sign-off at the end of each review.

Design reviews should highlight good design practices.
They should suggest ways to avoid known, design problems.

The design reviews should be considered as a dialogue that continues until all aspects of the design are complete.

Designers together with experts from both internal and external support groups will meet for design reviews, the design will be discussed and the use of correct design operations verified.

Reviews will take place at appropriate points in the development.

All ASIC suppliers will have their own design review procedure.

ASIC vendors will require a 'signed off' final design review before they will proceed further with manufacture of the design.

8.3 Design Stages

A possible scenario for the stages in a design and suitable points for design reviews is given in the following sections.

8.3.1 Pre System Design, Review 0

A management review should precede any ASIC design :-

*   System specification ;
*   ASIC procurement specification ;
*   Cad interface and design route ;
*   Software status ;
*   Circuit design considerations ;
*   Competence profile requirements of the designers ;
*   Define resources required and their availability ;
    Personnel, Computers, Cad, etc.
*   Timescales.

8.3.2 Pre Logic Design, Review 1

After choosing the basic form of a project and an ASIC supplier, the technical management must decide :-

*   Design route methodology to follow ;
*   The technology on which to map the design ;
*   Logic design details ;
*   Designer(s) ;
*   Test strategy ;
*   Package.

8.3.3   Post Logic Design, Review 2

After the logic design has been completed and before starting the physical design, - i.e. the layout. The following topics will be reviewed with the ASIC supplier :-

*   Logic design and verification ;
*   Test pattern generation and verification.

8.3.4   Post Layout Design, Review 3

After the completion of the physical design, the resimulation of the system with the addition of the loads due to the interconnection pattern will take place. This stage will cover :-

*   Layout ;
*   Post layout simulation and verification.

8.3.5   Post Project, Review 4

In order that the System House should obtain the maximum benefit from the experience gained on the overall project, it is necessary to discuss the following aspects :-

*   Identify from the design those parts that were good or bad ;
*   Improvements in methodology required ;
*   Improvement of CAD required ;
*   Improvement in aspects of support required ;
*   Bottlenecks ;
*   Strengths and weaknesses of the design team ;
*   Project success.

Chapter 9   Training

9.1   Introduction

Successful ASIC designs must have staff trained in :-

*   the basic operating language of the computing system ;
*   the basic manipulation of the CAD tools ;
*   a methodology, how to use the CAD tools effectively.

It will only be necessary for some designers to cover all tools.
This topic has been discussed in an earlier chapter 'Designers
Profile and Design Productivity'.

The training courses will need to be as practical as possible, with the
examples covering problems encountered in typical system designs.

In the methodology training, the expertise and experience of senior
members of the particular design group should be used to ensure
that the training covers the special needs of that group.

9.2   Training Personnel

There are three options for the personnel to run training courses.

Firstly, it could be sub-contracted to suitable external lecturers.

Secondly, all the training could be carried out by in-house staff.

Thirdly, could be a mixture of the first and second options.
Usually internal staff would cover the methodology.

The choice will depend upon :-

*   availability of suitable staff ;
*   availability of suitable accommodation ;
*   availability of computing power.

9.3   Computer Operating System

All users of the design system will need to have a minimum understanding of the computer operating system under which the CAD program is run.

As much of the operation will be hidden by the 'shell' of the CAD software little real knowledge is needed by general users.

It is necessary however for all users to have knowledge of :-

*   directory and file structure ;
*   file editing ;
*   impact on other users of particular operations ;
*   interfacing with other operating systems.

Support staff will need a thorough training as to all aspects of the operating system including system management.

9.4   CAD Tools

Normally, the initial training on the use of the newly acquired CAD tools will come from the software suppliers.
Additional training on these tools, may be a repeat of the initial courses for new staff. In the interests of economy or to customise for a particular need, courses may be specially prepared and run by the support staff.

Experience has shown that it is unlikely that one course can covered all aspects of the design tools and it may be more effective to use several short courses.

The training is run to map the design stages :-

*   System Description ;
*   System Simulation and Results Analysis ;
*   Synthesis to Netlist Description ;
*   Netlist Simulation and Results Analysis ;
*   Physical Design ;
*   Resimulation with Layout Effects ;
*   Checking Procedures.

The training will cover the aspects of the software programs described in the chapters covering those topics.

9.5   Methodology

The CAD tools on their own will not design a working ASIC but a disciplined approach to the use of the tools by a designer may.

The designing procedure should be based on experience and expertise in the understanding of the tools' strengths and limitations.

The chances of the ASIC being 'right first time' will greatly improve with this systematic approach to its design.

This disciplined and orderly approach to the use of the CAD tools is called **Methodology**.

Methodology should be based on the experience and design style of the System House.

The methodology should include :-

* a basic procedure, common to all designs ;
* special features for particular application ;
* design reviews at fixed points in the design sequence ;
  these should signed by a senior engineer or manager ;
  they should include checking for known good design practices ;
  they should include checking for common bad practices ;
* the methodology should be improved by design experience.

The content of the training course should be a blend of the use of the design tools and the methodology.

PART 3

Technical Details

Chapter 10 CAD

10.1 Introduction

Once again it must be emphasised that this chapter discusses :-

Computer **AIDED** Design

NOT Computer Design.

Designers must do the thinking and make all critical decisions.
They will understand the models used and their limitations.

The software will be left to carry out much of the routine work and to invoke the built-in checks that monitor the design.

Computers and the software programs that run on them, are now an integral part of IC design philosophy.

They are being used for most stages of the design, development, verification and manufacture of Silicon Chips.

This was not always the case, early attempts at Design Automation (DA) were tried but were far from successful because they were, probably, too ambitious for their time.

Some software design tools have been available for many years but they tended to be limited in application, slow in operation and unfriendly in their user interface.

A more meaningful and successful approach to a comprehensive CAD capability was developed to support an emerging design technique in the late '70s and early '80s.
Originally called Semi-Custom, it is now known as ASIC.

The ASIC design philosophy, for the first time in silicon design, laid a heavy emphasis on the expectation that the customer would receive 'right first time' chips no matter how complex.

This saw the implementation of :-

* controlled design procedures ;

* some restriction on design options ;
  only previously proven and characterised libraries of circuit functions could be used ;

* design for testability ;
  the testing of the prototypes was a major consideration of the design, from the very outset, beginning at the highest level of system specification.

It naturally lent itself to CAD techniques.

The objectives for using CAD were to :-

* Minimise design errors ;
  by building-in design checks where possible ;

* Increase design productivity ;
  by reducing the druggery and routine of the design by automatic procedures, particularly the later stages ;

* Facilitate design modifications ;
  by the use of good analytical tools ;
  and simple editing procedures.

The System Houses when undertaking the implementation of a 'System in Silicon' from the 'Specification to Working Chips' will find these factors are vitally important in improving the chances of first time success.

The Silicon Vendors in manufacturing, testing and supplying ASICs will depend equally upon the use of good CAD and CAM,
- Computer Aided Manufacturing.

Both will also require a good Design Methodology.
The major challenge for the CAD Vendors is to develop and supply a range of suitable, efficient tools.

These tools should have the following characteristics :-

* support all disciplines and technologies ;
* have an industry standard interface ;
* keep abreast of the requirements necessary for designing systems that reflect the continuing explosion in the complexity of ASICs ;
* support the users' design methodology ;
* be efficient in the use of computer resource ;
* be user friendly.

```
-------------------------------------------------
Remember, the design of a 'System in Silicon'
on an ASIC will depend on the use these tools.
-------------------------------------------------
And Success can only be measured by the supply
of ASICs that have been designed, manufactured
and supplied as fully-functional, tested parts
         in the right timeframe.
         ----------------------
```

Simply stated, the CAD should be available to help in the verification of a system at each and every stage of the design.

The designer should be able to use some or all of the tools available in the various suites of software programs.

The designer should be able to enter and leave the design procedure at many places. Provided that all critical checking routines are observed, the system should be flexible enough to meet local needs or design prejudices.

10.2   Requirements of CAD

The total CAD package for designing ASICs will, from the user viewpoint, resemble an iceberg, because much of the facility will be hidden below the surface.

The various programs in a software suite will fall into one of two main categories :-

those that will be used mainly by the designer ;
those that will be mainly invisible to the designer.

The designer will require visibility of :-

* System Description ;
* Detailed Design ;
* Simulation and Analysis ;
* Verification of the System Performance ;
  within its final operating environment ;
* Physical Layout ;
* Resimulation and Analysis, including Layout effects.

The programs not visible to the designer will, nevertheless, be crucial to the success of the final development stages before the manufacture of ASICs.

These will include software used to :-

* Run DRC - Design Rule Checks ;
* Run ERC - Electrical Rule Checks ;
* Run LVS - Layout versus Schematic checking ;
  thus ensuring the correlationship between physical design and system description ;
* Partition the layout into the data for each process stage ;
* Prepare that data for the Pattern Generator ;
* producing programs to drive the testers, ATEs.

10.3   CAD Tools

The CAD tools the designer will not use or usually will not see, are the province of the Silicon Vendor and will not be described.

The CAD Tools that will be covered, will need to be used at some stage of the Design Flow, from

System Specification to Tested Product.

These Tools will need to have Industry Standard interfaces,
        - e.g. **EDIF2, VHDL,**
Electronic Data Interchange Format, version 2.
This should allow transference of data to other design systems, where both hardware and software may be from different suppliers.

The software modules used for ASIC design will fall into one of three main areas.
The complete design will use each in turn.

> System Description
>
> Simulation and Analysis
>
> Layout - Physical Design.

10.3.1   System Description

The programs used for this function will depend upon expertise and experience of the System House and the complexity of the ASIC.

They will be able to support both :-

> Top Down and Bottom Up

design philosophies.
'Top Down' starts with a system specification and increases the detail as it progresses.
'Bottom Up' starts with the basic building blocks and increases the complexity as it progresses.

Some of the programs that could be used, are listed below.

* Schematic Capture ;
  drawing a circuit on a workstation ;
  at any hierarchical level ;
* Behavioural Description ;
  used at appropriate levels of abstraction ;
* Architectural Synthesis ;
  used to determine structure from behavioural description ;
* Logic Synthesis ;
  used to determine detailed netlist from higher level description ;
* Module Generators ;
  for ROM, RAM, PLA etc ;
* Silicon Compilers.

These will be used to input data at several levels of abstraction, the detail is described in the chapter 'Hierarchical Partitioning into Abstraction Levels'.

The purpose of these programs will be to help in the definition of the system being designed.

10.3.2   Simulation and Analysis

After confirmation of the system description, the designer must prove
its operation by using input stimuli to exercise it.
The program used will depend upon the level of system description.
Some of the following will be used :-

*   Simulators ;
    Circuit, Logic and Behavioural ;
*   Timing Verifier ;
    for certain types of system operation ;
*   Stimuli Generators ;
    for automatically providing the test vectors ;
*   Results Analysis ;
    for confirming the operation and checking for problems ;
*   Fault Simulator ;
    for confirming the signals as a potential test program ;
*   Test Analysis ;
*   Self Test ;
    Bist, JTAG, Scan Path, etc.

Firstly there will be the generation of the signals.
These signals will exercise the system in the simulator.
The results will be analysed for functionality and any possible system
difficulties, e.g. timing errors, glitches, races etc.
After successfully proving the functionality, the vectors will be
assessed for their use as the basis of a test program.
This will use the Fault simulator.

10.3.3   Layout - Physical Design

After the initial design has been completed, the physical picture of
the chip must be created. Some or all of the following programs aid
the successful completion of the task.

The first two :-

*   Floor Planner ;
*   Chip Size Estimator ;

will be used before the detailed design is complete.

The other programs will be used after the initial design stage to convert the functional blocks into their physical representation.
The shape, size and position will be used to create the picture of the chip as it will finally appear.

* Placer ;
  automatic and interactive, handling hierarchy ;
* Routing ;
  automatic and manual, handling hierarchy ;
* Track Load Extracter ;
  used to check the effect on critical paths of the additional capacitance presented by the interconnection pattern.

10.3.4   Additional Facilities

After the design has been completed but before the completion of the processing, many checks will be carried out.
These checks will be included to increase the probability of
'right first time' silicon. They will be used to produce good design practice and should include :-

* Design Rule Checker - DRC ;
* Electrical Rule Checker - ERC ;
* Layout Versus Schematic - LVS ;
* Power Estimator ;
* Power Calculation.

The first three programs will be used as checking procedures that the ASIC designer may not see or use.
They will normally be run by the ASIC supplier.
The last two programs will be used before and after layout to estimate and confirm the value of power that will be generated on the chip during forecast operation.
These calculations will ensure that the junction temperatures do not exceed the process specification due the combined effect of ambient operating temperature, rise due to the power dissipated on the chip and the thermal properties of the package.

10.4   Software Packages

Software Packages for the design of ASICs will be available mainly from two sources :-

* ASIC suppliers ;
* Independent Software Houses.

Each supplier's software will have advantages and disadvantages. These must be assessed in the light of the circumstances prevailing, at that moment in time, at the particular System House.

If the user expects all the CAD modules to be the best performers on the market, it is unlikely that they will be included in one company's CAD turnkey system.

The future for system design will lie in a flexible approach to the use of software, possibly based on a frame-work.

Provided that there is a consistent user interface, the designer will be able to run different modules of a program on different types of hardware within different operating systems.

The design manager must determine the design methodology on the currently supported softwares that will use the available design resources, both CPU and data storage, to their maximum efficiency and to the company's advantage.

10.4.1   ASIC Suppliers

These will normally be integrated design systems developed around the suppliers own silicon.

Many semiconductor companies will, however, support designs carried out on software other than that supplied by themselves, provided that an interface specification can be agreed.

The ASIC suppliers will have a different attitude towards their customers than the Independent Software Houses.
They will be anxious to establish long term, stable relationships with the successful customers that will offer the opportunity of supplying increasing volumes of silicon on a regular basis.

Their main product is the silicon not the software.

10.4.2   Independent Software Suppliers

This will be general purpose software capable of designing systems involving many technologies - ASICs, Standard parts, PCBs.

They will normally support the libraries of several ASIC Suppliers on their software.

The design flexibility will mean that the built-in checking procedures can only be those common to ALL the technologies they support.

As the business of these companies is selling software, their products tend to change often as they introduce new features.
This is intended to improve their overall, market appeal.

10.4.3   Turnkey Systems

A Design Software Package will consist of a number of modules.

Each module will perform one specific part of the total design.

Although interfacing with each other, the module usage and development will usually be separate.
It is possible for the various modules to change separately and not always in-step.

There will be a tendency for the offering from the suppliers to 'leap-frog' each other in the progress made on comparable modules.
Sometimes one supplier's module will be more advanced, sometimes another's.

This means that it is unlikely, that on a module by module basis, all the best performers on the market will be included in one CAD turnkey system.

It would be unwise for the ASIC design user to expect this to be so.

10.5   Future

The Design Pyramid shows that the majority of the labour used in a design will be consumed in the later stages of the design.

These stages, Gate level simulation, physical layout, resimulation including track loads, are to ensure the validity of the design against the silicon supplied.

Before they could be omitted it would be essential to build up confidence in the predicted performance and checking procedures available at earlier stages in the design.

Certain questions must be addressed.

* Will it possible to drop all verification at RTL/gate level ?
* Will it be possible to run timing, errors, glitch and fault analyses at a higher behavioural level?

If so, how much of the work currently undertaken in detail at the lower levels, can be moved up the hierarchy ?

Possible solutions would include :-

* More generators of parameterised functions ;
* Interactive verification during the layout phase.
  This would allow the analysis of the effect on the system performance of critical modules, including the layout, before the total layout been completed.
  This would allow the hierarchy to be maintained ;
* The introduction of more formal design methods.

10.6   Conclusions

The potential for improvements in system design using modern CAD programs is very great, provided they are used correctly.

Designers must learn to use the tools :-

* effectively and efficiently ;
* within a sound design methodology ;
* in the manner that meets their company's products ;
* in the manner that meets their company's design style.

Remember it needs CAD tools and a methodology.

Chapter 11   Simulation

11.1   Introduction

Simulation is the representation of the functional and dynamic behaviour of an Electronic System on a digital computer.

The model accuracies will determine the correlation achieved.

Several types of Simulator find application in different areas of investigation and design.

1   Simulators that work at the electronic behavioural level

Using designer defined functions, including high levels of abstraction, analysing the response before design of the system is completely defined.

2   Simulators that work at the electronic function level

Using predefined blocks, analysing the response to input stimuli with defined levels - Nand Gates, Nor Gates, Flip-Flops, etc.

3   Simulators that work at the discrete component level

These deal with equations that relate Voltage, Current and Time

Using 'lumped' values, analysing the performance of the components - Diodes, Transistors, Resistors, Capacitors, etc.

4   Simulators that work at the basic material level

These deal with equations that relate Voltage, Current and Time.

Using distributed values, analysing the performance of fundamental components - PN junctions, Transistor Action, Resistance, Capacitance, etc.

Simulation should enable a designer to develop and analyse a circuit, more quickly and cheaply than other methods.
It should increase the confidence that when the system is made in silicon, it will function correctly the first time of manufacture.
The simulation should confirm that the performance fulfills the requirements of the entire system specification, not just those of the ASIC, i.e. including the delays at all interfaces.

It should be remembered that the simulation models do not - and possibly can never - fully represent the entire physical reality.
Models can only be an approximation of the real world, the limitations must be clearly understood.

For efficient use of designers' time, the majority of the simulation should take place at the highest possible hierarchical level.
Experience has shown that to a first approximation, simulation time increases by an order of magnitude for each level down the hierarchy from algorithmic to geometric.

For a full understanding of and a high confidence in the system's performance, simulation should cover as many functional options and dynamic performance variations as possible.
Examples at lower levels should include :-

Maximum Delays *	Typical Delays *	Minimum Delays *	Supply Voltage	Temperature

This will allow the investigation of many worstcase situations.

* It is assumed that ALL gates will vary together as they are diffused on one chip. - i.e. All fast, all slow or all typical.

11.2   Simulation Modes

Before discussing simulation modes, it is necessary to distinguish between time measured in CPU seconds and elapsed time.

CPU time	Time in seconds, the measure of actual usage of the computer on this task
Elapsed time	Time between the operation beginning and ending

CPU stands for Central Processor Unit.

There are two common ways of operating simulators :-

interactively   and   batch mode.

In a single user computer system there will be little to choose between interactive and batch operation.
In a multi-user system, interactive simulations may generally be slower in terms of CPU time than the batch-oriented ones but as they make incremental debugging of the system much easier and the elapsed time is usually shorter, they are frequently used.

11.3   Simulator Types

There are several, fundamentally different types of simulators for evaluating electronic systems :-

      Digital   Analogue   Mixed Analogue and Digital Mode

They will be described in the following sections.

Figure 11–1 :   Simulator Interface

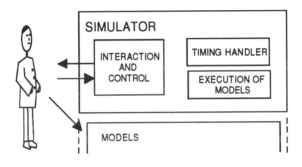

11.3.1   Simulator Construction

Today's Simulators must have the following capabilities :-

* Access to libraries containing electronic models, including physical parameters, functionality and timing data ;
* Ability to describe the system to be simulated ; accepting user-written models and stimuli, at any level of abstraction ;
* Compiler to convert the data into a suitable format ;
* Test vector assembler and event scheduler ;
* Control interface to interpret the commands and the results.

11.3.2   Digital Simulation

The majority of large digital systems use event driven simulation.

Event driven simulators may not be continuously active.
They will re-evaluate the output of a functional block when an input changes state. Any change on a function's outputs will propagate along internal connections to other modules in the system, causing re-evaluation of the state of all connected blocks.
Re-evaluation triggered by the initial event will continue until all activity has ceased.
When another input change takes place, it will give rise to more activity. Several evaluations may occur in parallel.

An event is defined as a net changing state.
A net is the input or output to a function.

The states will vary with the simulator being used.

In general, they will be related to :-

Logic 0,	Logic 1,
Unknown - X,	High Impedance - Z,

Timing will be related to a change between any two of the above stable states when a transient condition occurs.

In many commercial simulators, the logic states will also be given weighting, that is how strongly or otherwise that state is represented.

These weightings will help resolve the circuit ambiguities that can occur due to the presence of the Unknown or High Impedance states on the inputs of some logic gates.

The increasing number of available states, of necessity, will complicate the design system, increase the use of computer power required and increase the amount of data stored.

11.3.3   Analogue Simulation

Will fall into categories 3 or 4, defined in the Introduction above.

This mechanism for simulation is often called continuous simulation.

The inputs will be continuously changing, with small increments of voltage, current and/or time being applied to the device under test.

This method will use abstraction levels where physical processes control the circuit.
Typically the behaviour will be described by differential equations.
The continously changing inputs of voltage, current or time will require the simulator to solve these equations by iterations.

11.3.4   Mixed Mode Simulation

This means that several mechanisms can be used in the same simulation - e.g. event driven and continous mode together.

In other words, combined digital and analogue simulation.

The two simulators run separately but have a defined interface.

When the outputs of the analogue simulator reach specified values, they trigger an event based response in the digital simulator.

Similarly, a conversion of digital output values is used to drive the analogue simulator inputs.

11.3.5   Multi-Level Simulation

A multi-level simulator describes modules and signals at several levels of abstraction and allows them to be simulated simultaneously. As signals at different abstraction levels can be of different types, it is necessary to use signal translators for inter-level communication in the simulator.
An example of different signal types in use together is :-

> commands and data at one level
> bit-oriented signals at another level.

11.4   Fault simulation

Fault simulation is a technique used to verify the input patterns as a possible Test Program.

The normal procedure is to take the vectors, that prove the circuit functions correctly, as a starting point.

The test vectors of the simulation are used to detect artificial faults that are introduced by the computer when it is run in Fault Mode.

The industry standard fault model is :-

> Stuck-at-0, Stuck-at-1

A model is created where the nets external to the library functions, are connected to each of the power supplies in turn and so cannot change state during that simulation. i.e. only those interconnections that are unique to the specific design, not the connections internal to the library models. These special connections, known as faults, remain fixed at either Logic 0 and Logic 1.
Now the simulator will be run with these known faults included in the circuit interconnection list.
The 'good' and 'faulty' circuits will be simulated together.
The outputs will be monitored to detect differences in the waveforms between the good circuits and the faulty ones.

If the waveforms at the outputs have become different then the faults have been detected.
The simulator will then issue a report on the percentage of all faults that have been detected by the particular input vectors.

Hence the expression 'Percentage Fault Coverage' which is used as a measure of the effectiveness of the input vectors.
See also the chapter on 'Testability'.

On large circuits with long test patterns, fault simulation will consume large amounts of CPU time and data storage space.

In modern simulators, a concurrent fault simulation mode is used.
Models of all faults are created and simulated at once, fault collapsing techniques are used to reduce the amount of computing power needed but it can still be large !

11.4.1   Hardware Accelerators

Hardware accelerators are used to reduce the long simulation times dramatically. Typically accelerators run at more than a 100 times faster than conventional design tools.
Hardware accelerators support top down hierarchical design verification with both functional and behavioural representations.

The following models must be supplied :-

> Behavioural models
> Logic models
> Timing models
> Fault models
> Power Dissipation models

11.5   VHDL

In general in the past, simulators were propriety and rarely could circuits designed on one be readily transferred to another one.

This meant that System Designers were locked into one Silicon Vendor unless they were prepared to pay two licencing fees and two development costs on each design.

Some standard interface should be made available.

Many companies, particularly military contractors, may be required to supply products over long periods. The systems are supplied with currently available technology. With the rapid changes in silicon, this technology may become obsolete quickly. The system must be supplied with obsolete and difficult to obtain technology or incur the cost of a re-development.

Behavioural Specification of a system can offer a long term solution.

The ability to specify a system in a high level language will mean that the system specification can remain constant whilst the components go through several generations of upgrading technology.
If the language of the specification is an industry standard, it makes the supply of components of various scales of integration and from different manufacturers possible.

The overall system definition does not need to change over the lifetime of the project but the technology in which it is realised and the supplier can be changed many times.
The system can have a number of sources, each originating in a high level definition that can be read by each supplier's own software.

VHDL may cover these needs.

VHDL is the acronym for the :-

VHSIC **H**ardware **D**escription **L**anguage.

VHDL was sponsored by the **US DOD.**

It is now **IEEE standard 1076-1987.**

VHSIC is the acronym for another American Government sponsored development and this time it stands for :-

Very **H**igh **S**peed **I**ntegrated **C**ircuit .

11.5.1   The Purpose of Behavioural Language

The behavioural language provides a convenient way of modelling system performance instead of describing a hardware component in terms of its constituent, structural parts.

It can offer a number of advantages :-

* Allows a system to be modelled and simulated before its detailed design has been completed ;
  useful for evaluation in the early stages of design ;
  many ways of achieving the same system may be tried ;
* Allows an existing component to be modelled from its specification, with little knowledge of its internal structure ;
* Allows fast simulation models to be written ;
  by ignoring irrelevant internal detail ;
* The simulation of the complete system including its real environment may now be possible where before only parts of the system could be covered at any one time.

11.5.2   Who Should Use the Behavioural Language ?

Experienced designers who are working at system level in the early stages of a design.

Any ASIC designers needing to deal with large systems, should be able to model and simulate not only the specific IC they are working on but also the entire system, with all the interfaces between components, in a top down manner using the features provided.

Very fast simulation could be available at the more abstract levels of description, allowing the possible investigation of several solutions.

11.6   Conclusions

Simulation times do not increase linearly with the increase in the complexity of design. Their increase in simulation time appears to be somewhere between a square law and a cube law.
The spread is due to the difference in designing structured circuits, which can use timesaving facilities for repetitive functions and random circuits which cannot.

The same problem will also apply to the use of computer resources.

The very large systems designs now being considered may be impossible to verify at the lower abstraction levels.

If the continuing increase system complexity is to be managed, the current design systems must improve their performance by orders of magnitude
Possible solutions are :-

* hardware accelerators, preferably embedded in the standard design tools ;
* system designers simulating at the highest possible abstraction level for the majority of the design ;
* standard functional blocks written in high level language ;
* better integration of mixed mode simulation.

Chapter 12   Hierarchical Partitioning into Abstraction Levels

12.1   Introduction

This chapter deals with the partitioning of large electronic systems.
The systems will be divided into functional blocks.
These partitioned blocks will be designed individually and may use any of the methods of system definition described.

These methods will vary from :-

>   simple logic functions with their netlist interconnection ;

to

>   complex behavioural definitions with no internal details.

Many different terms are used for the hierarchical levels involved in the ASIC design flow.
This chapter will discuss following items :-

*   Definition of hierarchical levels ;
*   Alternative names for these levels ;
*   Typical design activities at each level ;
*   Different signal types and abstractions.

The major reasons for introducing hierarchy will include :-

*   Making the design manageable ;
*   Reducing the simulation time ;
*   Partitioning the design work between different designers.

The unambiguous specification of a complex ASIC design may require that it can be partitioned into several sub-specifications.
This may also improve the definition of the overall system.

Partitioning into several hierarchical levels of abstraction may be necessary to facilitate parallel design.
Each designer will have control over a section of the system.
Another advantage may be to reduce simulation time.
Much of the time the designers will work independently, in parallel, ultimately coming together for the whole design.

For small ASICs with complexities under 1000 gates there may be no benefit from the introduction of hierarchy.

The number of hierarchical levels needed will depend upon :-

*   The type of ASIC ;
*   The structure of the ASIC ;
*   The complexity of the ASIC.

Figure 12–1:   The Four Hierarchical Levels of VIDM

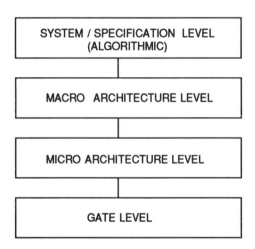

One method of designing complex systems,

VIDM, Vertically Integrated Design Method,

has been described in the chapter 'Design Methodology'.

In general, the more complex an ASIC, the greater the need will be for hierarchical levels.
The number of levels will depend upon the type of ASIC being used and the design structure.

Typically at each hierarchical step, one part will be replaced by between 10 and 100 simpler parts.

The four-level hierarchical method described in this book will be appropriate for ASICs to, at least, 100000+ used gates.

Unnecessary use of levels of hierarchy may create problems.

One example of the problems likely to occur in multi-level systems will be associated with interconnections of individual functions.
Unambiguous reference to any part of the system will require the names to reflect the depth of the hierarchy.
This will result in the necessity for a large number of characters in both net and block names.
Remembering and reproducing long names will be more prone to error and will also generate more data to be stored. This will make the comparison of results from simulations obtained at different levels of abstraction more difficult.

12.2   System Level Definitions

Each System House will introduce their own definitions of the terminology that they use in the specification and description of complex electronic systems.

This can lead to ambiguity.

In order that misunderstanding is reduced to a minimum, the terminology and format for many of these system descriptions, will be given in this chapter and used in the rest of the book.
The following sections define the meaning that the authors use. Where the expressions in common usage in the electronics industry are unambiguous, they will be used. In other cases, any uncertainty will be removed by defining the expressions.

Figure 12–2:   Hierarchical Relationships

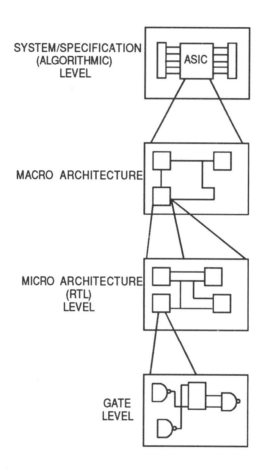

SYSTEM/SPECIFICATION
(ALGORITHMIC)
LEVEL

MACRO ARCHITECTURE

MICRO ARCHITECTURE
(RTL)
LEVEL

GATE
LEVEL

12.2.1   Specification or Algorithmic Level

Other names used at this level may be :-

Behavioural level, Highest System level, System level

The major objective of this level will be to write a specification for the complete ASIC.

Figure 12–3:   ASIC Definition

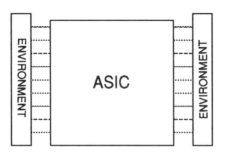

The correctness of the total system will be verified by simulating the proposed ASIC in its relevant environment, so as to include the interfaces which have frequently been a source of error.

This will be the highest level of abstraction used by the design team and the behaviour of the ASIC should be specified in a High Level or Hardware Description Language such as VHDL, described in the chapter 'Simulation'.

The environment, in which the ASIC under design will be required to operate, may include other ASICs of various technologies, standard components or abstract representations, like a radio channel.

The relevant environment should also be considered when generating the stimuli that will be used for verification of the ASIC. This should be the case at both this level and, indeed, at any lower levels of abstraction.

The signals used at this level may be abstract, including such types as data and commands. Any timing associated with the signals at this level will be only modelled to help determine that the system is functioning correctly.

e.g. clock signals may be modelled as enabled or disabled.

This level should be characterised by its compact description of the system function and very abstract signals.

Figure 12–4:   System/Specification Level

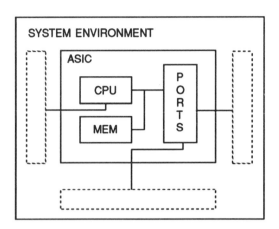

Typical design activities at this level will include :-

1   determining the functionality and performance of the ASIC.

2   deciding the extent of the environment that must be included to allow a verification of the ASIC's overall performance.

3   specifying the behaviour of the ASIC in VHDL.

4   simulating and verifying the ASIC in its environment.

5   considering the necessary testability strategy.

12.2.2   Macro Architecture Level

Alternative name used may be :-

Architectural level.

The Architecture of an Electronic Function will be a plan of the form and order in which the system may be realised. The ASIC will be partitioned into well-defined blocks, using the optimum architecture. For efficient use of the design effort, each block should be of a manageable size and complexity for one designer.

Figure 12–5:   Macro Architecture Level

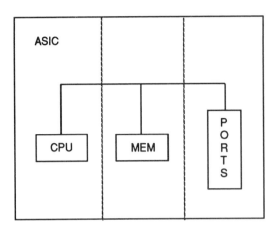

Now each block can be designed independently and in parallel. The signals used at this level may still be abstract types but it will be unnecessary to consider such details as - timing, clock skew, etc. The overall performance will be verified by simulation.

Typical design activities at this level will be :-

1   determining the architecture of the ASIC by partitioning the total behaviour into several sub-behavioural blocks where their functions and communication can be specified.

2   laying down the rules for communication and protocol between the different blocks on the ASIC.

3   coding the behaviour of each block in VHDL.

4   verifing the performance of each block individually and communication with other blocks and IOs at this level.

5   verifying the system against higher level specification by the use of multi-level simulation.

6   considering the layout floorplan.

7   selecting the test methodology.

12.2.3   Micro Architecture Level

Alternative names may be :-

> RTL, Logic level, Synthesisable level.

At this level, the individual functional blocks will start taking their final form. Many will be of the register type. One of the major objective will be to perform scheduling, resource allocation and sharing.

Figure 12–6:   Micro Architecture Level

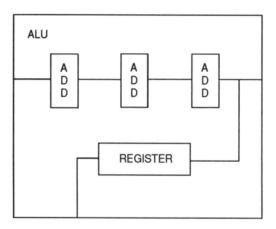

Design productivity will be increased here by use of synthesis.

The description at this level should still be the highest possible level of abstraction acceptable by the synthesis tool.

The design will be translated into the macro functions, RAM, and ROM that are available from ASIC vendors.

Signals will become bit-oriented at this level and timing become clock-cycle related.

Detailed test methods will be selected.

Test methodologies like Scan Path will be introduced, if necessary.

Typical design activities at this level will be :-

1   selecting and dividing into smaller blocks, each with its

     connectivity must be coded, using synthesisable VHDL ;
     description may be in the form of an instruction set ;

2   introducing sequensers and controllers using
     microcode or state diagrams as inputs ;

3   scheduling and resource allocation ;

4   sharing performed explicitly (manually) or
     implicitly by a synthesis tool ;

5   defining dataflow and controls ;
     including parallel operation, pipelining etc ;

6   building datapath registers and arithmetic units ;

7   verifying blocks individually and their communication
     with other blocks and IOs at this level ;
     using multi-level simulation to verify against
     higher level specification ;

8   physically interconnecting blocks, IOs and bonding pads ;

9   planning physical layout ;

10   implementing test, either manually or automatically.

12.2.4   Gate Level

Alternative names used may be :-

> Logic level, Lowest level, Primitive cell level

At this level everything will be implemented in gates, RAMs etc. using a library from the chosen Silicon Vendor.

The design work will be completed including detailed simulation, fault simulation, layout and resimulation.

At this level, logic synthesis and Automatic Test Pattern Generation - ATPG, will increase productivity dramatically.

Timing representation will change from clock-related to absolute. Bit-oriented Signals i.e. individual signals will become necessary.

Figure 12–7:   Gate Level

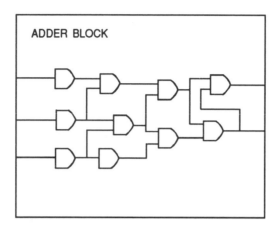

ADDER BLOCK

Typical design activities at this level will include :-

1   implementing the gates manually or automatically ;
     including synthesis from behavioural descriptions,
     boolean equations or truth table ;

2   simulating the logic in detail ;
     including timing, glitch, race and power analysis ;

3   generating test vectors ;
     both manually and automatically ;

4   analysing fault simulation ;

5   floorplanning and physical layout ;
     both manually and automatically ;

6   post layout verification of simulation results ;
     will include trackloads from the layout.

7   power consumption calculation ;

8   verifying the system against higher level specification ;
     by the use of multi-level simulation.

12.3   Other Definitions

As described in a previous chapter, the term **Gates** is the most imprecisely used word in ASIC technology.

In Gate Arrays, 'gate' frequently refers to :-

* basic electronic functions - Nand, Nor, Exor etc ;
* size of a Gate Array - 2040, 100,000, etc ;
* group of unconnected transistors ;
* group of transistors connected to form a function ;
* method of relating system complexity - 5000 Gates equivalent ; meaning the system is equivalent to 5000 2-input Nand Gates.

Semiconductor Suppliers give their own names to :-

* groups of unconnected transistors ;
* groups of transistors connected to form basic logic blocks ;
* groups of transistors connected as larger, standard blocks ;
* the special blocks built from the above blocks.

The following sections have been included in an attempt to remove some of the ambiguity that exists in terminology used in this subject and to produce a consistent text.

12.3.1   Algorithm

The description in mathematical language of the solution or solutions of a particular problem.
Initially it will not normally include any timing data.

12.3.2   Behaviour

Behaviour describes functionality of a module -
i.e. what the function is required to do.

It is from an external view only.

A 'Black Box' description, giving details of external connections only, what function(s) is (are) performed and the timing relationships.

At each level in a system, the behaviour is a description of the functionality that must be solved.

It is the interface at each level in the system.

12.3.3   Floorplanning

Floorplanning is the division of a system or part system in to functional blocks.
The size and shape of the blocks will be approximately determined by the contents but no details will be given.
It is used for investigating the possible physical relationships between the blocks.

A better expression is FUNCTIONAL PARTITIONING.

Functions are realised in blocks and/or components.
The interconnection is not necessarily specified.

The dataflow and controls will be defined - parallel operation, pipelining etc.

Parts of the description may be in the form of an instruction set.

12.3.4   Functionality

Similar to BEHAVIOUR but is more specific.
It describes modules that can be realised in different formats.

The final construction will vary according to the specific application.
The details will change for low power, high speed, small area etc.

12.3.5   Structure

Structure describes how the assigned resources are connected and how they communicate with each other - i.e. the manner in which the function is to be realised in this specific case.

Structure will detail the number of Datapaths and method of control.

12.3.6   Array Element

Smallest grouping of transistors in the basic technology.

In CMOS Technology

the most common format is 2 P-type and 2 N-type transistors.

Sometimes the group may be 6 transistors.

12.3.7   Cell

One or more array elements connected to form the regularly-used, basic logic blocks.

These represent the 'core' library of a particular technology.

They have fixed interconnection patterns and will have both simulated and hardware characterised performance data.

12.3.7.1   Macro

Next level of commonly used electronic functions.

Constructed from cells.

Can be 'hard' macros - i.e. fixed geometry including interconnection - producing repeatable dynamic performance.

Can be 'soft' macros - i.e. variable geometry within the constraints of the cell shapes. The dynamic performance requires checking after placement and routing to allow for the possible differences caused by the variations in routing. The system will be described in a mixture of the electronic functions given below.

Chapter 13   Synthesis

13.1   Introduction

In general, the term
>'to Synthesise'

means
>'to build up or create from separate elements
>a coherent or connected whole'

'Synthesisers', in the jargon of the Electronics Industry, are used in a number of product areas -   e.g. in RF Tuning.

In the context of this book, 'to Synthesise' will mean -
>to automatically resolve the specification of
>an electronic system from one level of its hierarchy
>to a lower level in the hierarchy, with the inclusion
>of more detailed circuit information.

Synthesis has been taken as a synonym for Logic Synthesis, i.e. a set of tools that will generate gate-level netlist structures only, from Boolean eqations and truth tables.

It will be necessary, therefore, when using the more general definition of Synthesis, to differentiate between the various modes, by using qualifiers, namely :-

>Architectural ; RTL ; Logic ; Layout ; Test.
>e.g Architectural Synthesis.

These terms have been defined in the previous chapter - 'Hierarchical Partitioning into Abstraction Levels".

Synthesis, which works as a three stage process, will be used for reducing the problems and much of the manual work associated with the design of complex ASICs. The three stages are :-

>Translation ; Optimisation ; Mapping

Figure 13–1:   The Levels of Synthesis

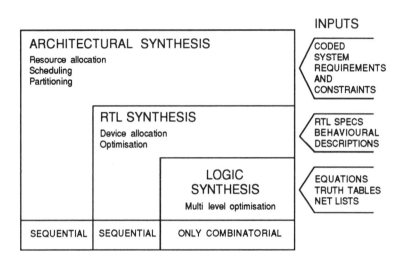

Among its main objectives will be :-

to define the performance of the circuits ;
to increase design productivity ;
to make the design, 'technology independent'
to reduce the number of errors made while designing.

There is a widely held view that the use of synthesis will permit the use of lower calibre designers. Experience has shown if synthesis is used the degree of competence will be reflected in the quality of the design. Controlling the inputs to the synthesiser and operating any options available will particularly show up any lack of expertise.

This chapter will deal mainly with Behavioural and Logic Synthesis.

Silicon Compilation can be classified in this group.

The differences in the formats of the inputs and outputs will depend upon where in the hierarchy the synthesis takes place.
See Figure 13–1

As already stated, it will be possible to use synthesis at

Architectural, Register Transfer (RTL) and Logic Levels. Synthesis at any level may offer a number of solutions.

Two interpretations of Behavioural Synthesis could be :-

1	Input	-	Behavioural Language Description
	Output	-	Behavioural Language Description but more detailed

and

2	Input	-	Behavioural Language Description
	Output	-	Structural Language Description

13.2   The Synthesis Options

One of the following forms of synthesis will imply the automatic creation of system data, starting from the named level of abstraction and moving down to one of the lower levels :-

Architectural	-	System Level
RTL	-	Register Transfer Level
Logic	-	Gate Level
Layout	-	Physical
Test		

Typical format options for each level of abstraction are given below.

Architectural Synthesis   Architectural Synthesis

1	Input, Behavioural	-	Output, Behavioural
2	Input, Behavioural	-	Output, Gate level netlist optimised

RTL Synthesis

1	Input, Behavioural	-	Output, Gate level netlist optimised

Logic Synthesis

1	Input, Boolean, Table or Netlist	-	Output, Gate level netlist optimised

Dedicated datapath synthesisers or module generators do exist. They can take user-specified parameters - e.g. number of bits, number of words - and, typically, generate :-

adders, registers, counters, multiplexers, decoders, comparators and other types of module required in a datapath.

They are easy to use and quickly produce a result but are limited in their effective applications.

Architectural, RTL and Logic Synthesisers should be looked upon as bridges between the system specification and the silicon.

They can be used for generating gate level netlists from different technologies starting from one common specification.

The designer can change the library used according to performance requirements of the design.
They could be used for generating gate level netlists for automatically mapping an ASIC design onto different Silicon Vendors technology libraries, for performance requirements of the design and/or a competitive situation on development and unit costs.

13.3   Implications of using Synthesis

Before a System House can use this type of design tool, the designers - and, to a lesser extent, their managers ! - must understand the implications of the input and output formats on which the various synthesisers can operate and any possible pitfalls.
Although these tools only appear to have advantages, unless the limitations of their techniques are fully understood, their effect on performance of any design and its timescales may result in major setbacks to the overall project.

The problems associated with the use of synthesis normally arise at the layout and test stages.

Experience has shown that most synthesis tools have not been developed with the effects on layout as a major input.
This has been particularly noticeable when they have performed optimisation on area and critical paths.
In the case of area optimisation, the most tools only consider the number of equivalent gates being used. They should consider both gates and their required routing area. Often the tools will produce outputs that use one gate with many inputs but only one logic level. Such optimisation has a tendency to require more routing area than multi-level logic with gates having fewer inputs.
The overall effect may be to produce a result that finally turns out to be too costly in area.

Another problem associated with synthesised structures, can be a loss of traceability in the automatically produced schematics,
i.e.   the new netlist can become difficult to relate specifically to the existing system.   This may be especially true with large

functional blocks where test analysis and manual layout, particularly placement, are likely to be especially troublesome.

Traceability could be maintained if synthesis was restricted to relatively small blocks but this may offset the productivity advantages predicted for the use of synthesis techniques.

Figure 13–2:   Synthesis Productivity

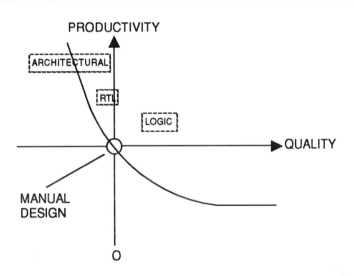

The curve in Figure 13–2 shows the relationship between design productivity and design integrity if synthesis was used at various levels of abstraction, taking manual design as reference.

Productivity has been taken to be a measure of the rate of doing a design, in gates per man-day.

Quality has been shown as a measure of what was required against what has been achieved in terms of system specification integrity.

High abstraction level definition will be quicker but less accurate. Lower level definition will be slower but more accurate.

There will be requirement for training and 'learning by doing' experience until the designers feel confident in use of these tools.

13.4   Evolution of Synthesis

Synthesis is part of a 'Top Down' design system.

By a strange irony, the evolution of Synthesis Technology followed the 'Bottom Up' approach.

The first Synthesisers were used to improve combinatorial gate level systems by processing with the Synthesis tool according to a critical input parameter.

Typically the system was :-

* modified to minimise logic used ;
* modified to minimise delay on critical paths ;
* modified to minimise overall area ;
* compared with an existing system for functionality.

The second generation was able to synthesise :-

> from Behavioural Description Language
> into a Gate Level Description.

13.5   Future Trends

Synthesisers are emerging that can work from higher levels of abstraction, such as

> Architecture and/or Algorithmic Level.

They are, at present at an early stage in their evolution and are far from 'robust'. They are expected so show dramatic improvements in the near future.

Synthesis as a whole when used in design of electronic systems, is a relatively new technique and as such has great potential.

The judgement on these improvements will depend very much upon how the currently viewed limitations are tackled.

There will be a growing need for compatibility with and better integration of the synthesis tools within other ASIC design tools.
Designs will need to move across both hardware platforms and between software programs supplied by different vendors.
This will enable the System House designers to take a more flexible approach during the design stage.

It should become possible to analyse more quickly any changes to input specifications. This should permit several options of the design based on input data from different technology libraries to be analysed. Before the final decision on the system implementation has to be made, the possibility of overall system improvements may be studied.

Changing the way that write a system specification from their traditional method, to using a high level language format will be a major step for some companies.
There will be a need change their current working practice because of the potential advantages that it will offer but there will be an initial reluctance to overcome.
After a period the advantages should become apparent.
They will need to build up an understanding of, and a confidence in, the correlation between the written VHDL code and the output from the synthesiser,

13.6   Synthesiser Operation

13.6.1   Introduction

Generally, a Synthesiser will perform three tasks.

This is illustrated in Figure 13–3.

Firstly the synthesiser will resolve the system from one level of abstraction to the next level lower, building the structure of the system from the higher level description
This is called translation or conversion.

Secondly, it will optimise the structure, whilst remaining at the same level. The new system will meet a particular input parameter or parameters, including the removal of any redundancy.
This has become an important feature.

Thirdly, from an appropriate level, it will map the current system description to the chosen Silicon Vendor's technology library.

The process of synthesis will normally be a continuous operation.
It can however be a two or more stage procedure if the designer chooses to monitor the output of each operation.
The output structure will include the attributes requested.

Figure 13–3:   Synthesiser Operation

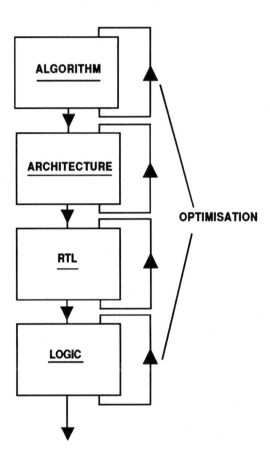

13.6.2   Architectural Synthesis

Typical activities at this level will include :-

*   System partitioning and Block interfacing ;
*   Pipelining ;
*   Resource allocation ;
*   Resource sharing ;
*   Test strategy decisions.

The system, coded in VHDL Behaviour Language (or something suitable) will be translated into the chosen architecture.
After the behavioural description has been translated then architectural optimisation can start.

Other examples of activities at this stage will be :-

* looking for shareable multiple arithmetic blocks ;
* looking for logic operations that can be performed on a single but more efficient block ;
* rescheduling operations.

Benefits

Working at this high level will allow an analysis of the trade-offs available within the specific design.
It will make the undertaking of complex designs more manageable.
It will support design exploration and innovation.
It will encourage the re-use of proven functional blocks.
The company's design productivity will be enhanced.
The 'right first time' success rate will be improved.

Experience at present suggests that synthesis from higher levels will only be effective on certain architectures. When used with other, more general architectures, synthesisers may be less successful. They may require considerable manual help to achieve a totally satisfactory result.

13.6.3   RTL Synthesis

Typical activities at this level will include :-

* Resource sharing ;
* Sequential optimisation ;
* Logic optimisation ;
* State machine synthesis ;
* Test overlay.

At this level some decisions about the style of the design will have already been made. The coded version of the function will be much closer to a particular hardware implementation than it was at the architectural level. The clocking strategy, synchronisation, interfaces and dataflow must be considered now. The format of the circuit will be determined by the use of different forms of a high-level coding language such as VHDL.

13.6.4   Logic Synthesis

Typical activities at this level will include :-

* Logic gate-level optimisation, two or multi-level ;
* Redundancy removal ;
* Technology mapping. ;
* Pad allocation.

The synthesiser may be guided by user constraints.

Logic synthesis will translate equations and truth tables into optimised netlists with a minimum of redundancy.
So far the procedure has been found to give the best results on combinatorial logic but tools capable of similar performance on sequential logic are emerging.

At Logic level, it will be possible to introduce extra circuitry.
These additional netlists can be created manually using schematic capture or by using previously designed functions.
It will then be possible to automatically optimise the additional netlists, both new and old, with the synthesiser, if needed.

Logic synthesis will be used map the netlist onto a specific technology library of a Silicon Vendor.

At this level of synthesis, it should be easier to port a design between different libraries and/or Silicon Vendors.

Typical input formats will include :-

Boolean equations ;
Netlists ;
Truth tables ;
State tables.

13.6.5   Test Synthesis

This activity will cover two main areas -

Testability and Test Synthesis.

Testability will cover making the design testable.

Test synthesis will cover test pattern generation and interfacing to the selected tester.

Testability Synthesis

This is part of the 'Design for Testability' procedure.
It will cover some or all of the following :-

* Test plan generation ;
* Test hardware insertion, (manual or automatic) ;
* Test rule checking ;
* Technology Libraries.

Test Plan Generation
The test strategy will be selected and a plan devised for dividing the ASIC into hierarchical blocks that will give easy access for testing.

Hardware Insertion
Additional circuitry, necessary for test will be designed into the ASIC either manually and/or automatically. This may include the generation of scan paths.

Rule Checking
This will check that the rules for testability have been correctly implemented.

Libraries
This will call up the required circuit blocks and test data from ASIC libraries being used.

13.6.5.1   Test Synthesis

This will deal with :-

* Test data generation ;
* Test output translation.

Test Data Generation
This will generate the bit-oriented pattern used for testing the ASIC. It may be possible to use Automatic Generation of the Test Pattern, ATPG.

Test Output Translation
This will translate data from previous section into a testpattern compatible with the language of the test machine to be used.

13.7   Performance

The use of automatic synthesis tools will result in less control over the detail in the ASIC design. If the circuit has not been created by the designer, the identity of each function and the interconnection pattern will have been given by the design tools, not the designer. Actually tracing the network through may prove a difficult task.
The designer may also find problems in determining whether the result from the synthesiser has given an optimum solution or not. Judgement on the result can only be based on experience.

At the present time, Synthesis has the reputation of generating more structure than would be produced by a good manual design.
The actual relationship between the results produced manually and automatically will depend on the nature of the system being designed and the technology being used but experience has shown that synthesisers produce results at least comparable with an average manual design and in a shorter time but not as good as the very best.

The apparent inefficiency may be partly offset by understanding the operation of the synthesiser and designing in a compatible mode.

Poor knowledge of the tools behaviour will normally result in poor designs, as is usual with computer AIDED design.

If high productivity, easy verification, compact designs and good dynamic performance are to be achieved with the use of synthesis, it will require close linking between different tools.

Synthesis at the Architecture, RTL and Logic levels must be coupled with the Functional design and Layout tools.

Judging the efficacy of the various options available will be very difficult. It will only be possible with experience.

13.7.1   Optimisation

Normally this is an operation that can be carried out efficiently only after a designer has gained experience.

The most commonly investigated variables are

area and critical path delay

The synthesiser will investigate the circuit configuration in terms of the input parameters chosen.

The output configurations may differ greatly from one another.

The results will need to be analysed in terms of previous design experience, particularly, the impact on the layout.

It is possible to produce a netlist which turns out to be extremely difficult to route.

13.8   Silicon Compiler

13.8.1   Introduction

The Silicon Compiler may be classified as a synthesiser with important, additional features.

Silicon compilers were very popular in the eighties but their weakness was in their lack of flexibility. They also suffer from a lack of controllability which limited their popularity amongst the users.

They will give the system designers great assistance provided that the system being designed is compatible with limitations imposed by the Silicon Compiler. Only a limited number of architectures do so.

On other systems, the Silicon Compiler will not be very effective, tending to be inefficient in the use of silicon area.

It is possible for a Silicon Compiler to complete the entire system design process automatically. This is provided that the system has a definition level that is within the capability of the software.

It was intended as the encapsulation of the best design knowledge and expertise into an integrated system that could carry out many of the routine operations of a system design. The drudgery of detailed work should become an automatic procedure in many designs.

This would allow the designers the freedom to concentrate on the creative work of system exploration.

Much of the drudgery and many of the time consuming operations have been concentrated at lower levels of the design triangle.

Theoretically, therefore the introduction of automatic procedures should result in a dramatic increase in design efficiency, measured in gates per man-day but there may be a price to pay for this reduction in design time :-

> larger ASIC area ;
> lower ASIC performance.

Figure 13–4:   Silicon Compilation

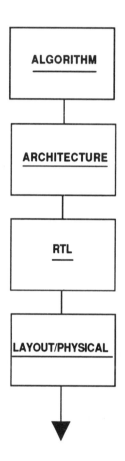

13.8.2   Operation of a Silicon Compiler

A silicon compiler will take input at a high level of description - algorithmic or architectural - and - automatically synthesise the chip. It should be able move from behavioural level description down through any necessary stages to the physical domain.
It should also be capable of automatically generating any necessary models for verification and analysis.

The principle of the silicon compiler is similar to that of software compilation. It take programs written in high level language and synthesises them to a lower-level, complex assembly language.
It follows the implementation rules in the compiler.
Typical inputs can include :-

*   Functional or Behavioural descriptions ;
*   Mathematical equations ;
*   Logical equations ;
*   State equations.

Only the first option will use the full capability of the tool.

The structures to be synthesised must be known to the compiler.
This is the reason why the operation of most existing Silicon Compilers is limited to certain types of architectures.

Typical structures in the architectures available will be :-

*   Data path ;
*   Control Blocks ;
*   Memories ;
*   Pads ;
*   Test Structures.

The ultimate Silicon Compiler will optimise and resolve :-

*   Algorithmic Description to Architecture Level ;
*   Architectural Description to Behaviour RTL Level ;
*   Behavioural RTL to Structural RTL level ;
*   RTL Structural to Logic Gate Level;
*   Test Overlay Synthesis ;
*   Physical Layout ;
*   All necessary design checks ;
    - LVS, DRC, ERC, etc ;
*   Generation of Slice Processing Data.

13.9   Conclusions

Synthesis should be used on large blocks, thousands of gates, for appreciable gains in productivity.

The overall design that includes the layout and testing, should be considered at the same time as synthesis.

The link between the system design and the silicon chip will occur at the Gate level synthesis when the libraries from the Silicon Vendors are introduced.

VHDL has emerged as the industry standard for System and Circuit Description. This should assist in making possible the transfer of system specifications between design tools.

Architectural and Logic synthesisers are continuing to encroach on the areas where Silicon Compilers were active. Experience shows an improvement in their effectiveness.

Synthesis should enhance design productivity and make complex ASICs more manageable. Also it should improve the chance of 'right first time'.

Synthesis should assist in the creation of re-usable functions.

When logic optimisation is carried out, the new generations of Synthesisers must include input parameters that are related to routing area. The solution given must be a compromise between optimum performance on the input parameters and a layout that can be routed.

An 'minimum area' solution without routing area is meaningless.

Only if the number of gates and the area allocated for routing are compatible is the design complete.

There must be a match between the design tools and library being used. i.e. the synthesisers must take advantage of all available functions and to use them efficiently.

Synthesisers do not yet optimise for power consumption, as this will be related to the activity of all nodes in the system implied by the input stimuli.

Chapter 14   Design for Testability

14.1   Introduction

The benefits offered by the increases in complexity of ASICs are partly offset by the increased problems and costs in making the products testable.

The discussion in this chapter deals mainly with the problems - and their solutions - of testing large digital or mainly digital circuits.

The total cost of testing ASICs can be very high.

This will be due to the cost of :-

> building Testability into the design ;
> generating test programs ;
> test time on the very expensive ATEs ;
> (Automatic Test Equipment).

In some cases, the test development may incur greater than 50% of the total development cost.

Only by careful and detailed planning from the beginning will a successful test program for a complex ASIC be developed.

This will mean starting at the specification stage.

The use of any special testing techniques must be considered before the start of the design cycle.

'Design for Testability' - DFT ,

will mainly use techniques that fall into two categories :-

> ad hoc ;

> structured approach.

The first technique will be used to resolve the testing problems associated with a particular circuit and the approach may not be generally applicable to other designs.

The second category covers techniques that will be generally applicable to sequential logic where test generation and verification may be inherently long. The approach will be to reduce this problem.

As the complexity increases, the functional testing of ASICs, even enhanced with additional test features, may still result in unacceptably long test times.

For the first time, the new methods may introduce the use of a test program that does NOT test the ASIC functionally in the manner in which it will be ultimately used.

Many companies will find this a major psychological barrier.
They will be asked to accept a product that has been subjected to statistical methods of testing for determining its acceptability.

The testing methods that are described below are an attempt to overcome these problems. The test mode will prove that large portions of the circuitry function, although that function will be different from that intended for the final system operation.
Statistically, therefore, there will be a high probability that the whole circuit will work.

Before undertaking any design incorporating these test methods it will be necessary for the design team to understand not only their application but also to understand all their limitations.

There will be no one approach that will solve all the test problems of any design team. It will be therefore necessary for the designer to have access to several different test methodologies.

The optimum solution for a given circuit can be chosen.

All designers will need to work to a set of carefully prepared testability rules. If followed, they should ease the development of a test program and, hence, the testing of the ASIC.

These guidelines will give to new designers an indication of the likely advantages to be obtained when choosing between the 'ad hoc' and the 'structured' approaches.

Table 14–1 gives a general comparison between the two.
As all parameters listed are likely to increase, the table indicates which of the two methods may be affected more.

Table 14–1:   Test Method Guidelines

Parameter	Ad Hoc Test	Structured Test
Chip Area	Smaller	Larger
Test Time	Shorter	Longer
Pin Numbers	Smaller	Larger
Performance	Higher	Lower
Design Time	Longer	Shorter
Design Cost	Higher	Lower

14.2   Testability

Remember

> an ASIC is a black box.
>
> Access is only possible to its Inputs and Outputs.

Testability is the capacity :-

> from the Inputs and Outputs available ;
> by a mixture of combinatorial and sequential signals ;
> to pass all the fully working circuits ;
> to reject all the faulty circuits.

Testability = Controllability + Observability

Controllability	that is the ability to force a known state on every internal connection from the declared inputs ;
Observability	that is the ability to observe known states on every internal connection from the declared outputs.

For very large circuits, it will be unlikely that a 100 % test program will be available based on economic grounds. (See next section.)

This will be due to a combination of :-

* the test time would be too long ;
* the cost of the extra silicon area would be too great ;
* the development of the test programme would take too long.

The test program must be viable economically. As 100 % may not be possible, it must also provide a worstcase, forecast of failures in product shipped that will be acceptable to both parties.

Controllability will be normally mandatory.

Observability will be a compromise between test time economy and acceptable fault coverage.

Most vendors will demand a fault coverage of at least 95 %.

The two parts of Testability taken together will give a very high probability that the devices delivered will be good.

14.2.1   Fault Simulation

Test programmes will be developed and assessed for their effectiveness by running the simulator in 'Fault Simulation' Mode.

Most Fault Simulators use a technique called :-

Stuck-at-0, Stuck-at-1.

After the simulator has been run in normal mode to prove the correct operation of the ASIC, it will be reprogrammed to run in a mode, where artificial faults have been introduced into the circuit.
When the test vectors are run the outputs of the ASIC are monitored and the results compared with the correctly operating circuit.

A 'detected' fault will cause the output pattern to change.
An 'undetected' fault will not cause the output pattern to change.
An 'undefined' fault will cause the output pattern to include X-states.

With 'unknown' states included, the outcome will not be clear.
Fortunately many of this type will be clearly detected by other routes.

The faults introduced into the circuit will cause the particular node to remain constant throughout the simulation.

The fixed values chosen, are Logic 0 and Logic 1.
Each node will be fixed in turn to each of these values.

This will normally result in long simulation runs and the use of enormous amounts of computer resource.
'Concurrent Simulation' can be employed to reduce the actual time needed. Instead of introducing faults one at a time, they will all be introduced at once and a technique called 'Faults Collapsing' will reduce these long simulations by removing any duplicated effects.

If all the artificial faults can be detected in an acceptable time, the test vectors would be said to achieve 100 % fault coverage.

Large circuits will not often achieve this figure.

The effectiveness will be measured as a percentage of this value.

14.3   Design for Testability

The System House should clearly define the specification interface itself and its Silicon Vendor.   Although, contractually the responsibility for producing a testable product may not be the designer's, the designer must understand that, if **DFT** is not practised from the beginning of the project, the result will be far from optimum.
Therefore, **DFT** should put the responsibility for good, efficient testing of the ASIC firmly with the designer.

Certain system functions are known to be difficult to test.

This is because :-

> they will require very long test patterns ;
> parts of the system will be very difficult to 'see' ;
> i.e. observable from the declared outputs.

The approaches to testing given below will explain possible ways of overcoming these difficulties.

DTF in general will increase the chip area used and may reduce the performance of the circuit.

It will be important to keep any extra circuitry included for testing purposes to a minimum.

Statistically, the more complex the additional test circuitry, the greater the chance that it, i.e. the test circuitry, will include a manufacturing fault that will cause the whole circuit to fail.
The forecast yields may be sufficiently lower that they will result in significantly increased costs.

The test circuitry may also be more difficult to check than the function as originally designed !

14.4   Testability Design Rules

The design manager should devise a set of rules that, for the design system being used, have proven to produce acceptable results. These rules should cover the testing of :-

*   random logic ;
*   PLAs ;
*   memories ;
*   embedded CPUs ;
*   mixed analogue/digital circuitry.

Remember the testing problem may increase in severity as whole systems become integrated.

Frequently the number of inputs and outputs available will decrease, making access more difficult.

14.4.1   Automatic Test Program Generation

ATPG may give acceptable results on combinatorial circuitry where the objective will be to generate a sequence of vectors that will report all possible faults inside the ASIC.

ATPG may give as little as 30 - 50 % fault coverage on sequential circuitry where the designer has a freehand with the system.

ATPG patterns are not recommended for debugging the initial design and functionality of an ASIC chip.

ATPG may be used to generate test vectors AFTER the design has been shown to perform according to the specification.

14.5   Ad Hoc Approach to DFT

The first approach will use a partitioning technique.

The circuit will be designed with the ability to isolate portions of the system specifically for test purposes.
This will make testing simpler and shorter.

The second approach will involve the introduction of an additional test point or points.

Either will only be possible if additional pins are available on the package unless the extra expense of a larger package can be tolerated on the project.

Both could also include the additional penalty of increased silicon usage, with the results discussed above.

The usual procedure for either Ad Hoc approach will be :-

> design the system ;
>
> prove its functionality with input vectors ;
>
> assess these vectors as a test program ;

with fault simulation ;

> improve test program with additional test vectors, circuitry and/or outputs.

Experienced designers will instinctively change the circuit to simplify the test pattern and to improve the fault coverage.

14.5.1   Alternative Strategies for Testing ASICs

The following approaches are available :-

* using the existing logic, add extra test vectors ;
* add extra circuitry, use existing test vectors ;
* add extra circuitry and test vectors.

Firstly

Take the functional patterns, analyse the fault simulation results, add extra test vectors to overcome the exposed weaknesses.

Produce a new test pattern to overcome the known deficiencies.

Secondly

Starting from an analysis of the initial simulation results, add extra circuitry to improve Observability of difficult nodes. This may only be possible if there are unused I/Os available.

Thirdly

Use a combination of the first and second methods.

14.6   Structured Approach to DFT

Most structured design techniques are based upon the concept that if the values on all memory elements in a design - flipflops or latches - can be controlled and observed then the test generation and, possibly, the fault simulation task can be reduced to that for a combinatorial logic network.
An input control signal will be used to switch all these elements from their normal mode of operation into a test mode.
This will make them more readily controllable and observable.

The memory elements will be strung together to form a shift register normally called a 'Scan Path'.
Values will be forced onto the elements of the path and then shifted around until they reach an output, where their values will be read.

This technique will enhance Controllability and Observability.
This augmentation of testing mode can make the control of inputs and internal states predictable and the internal behaviour of the circuit readily subject to analysis.

This technique has the obvious disadvantage of a serial format with a potential increase in test times.
Dividing up long scan chains into shorter lengths connecting them in parallel will be one method for reducing the scan test pattern.

These techniques can be used to introduce Self-Test Techniques.

There are many methods of introducing Self-Test to a system.
Some are proprietary and so require permission to use.

Most are dependent upon some form of 'Signature Analysis' and the use of a 'Scan Path'.

Forms of Self-Test include :-

*   BILBO - Built In Logic Block Observer ;
*   BIST   - Built In Self-Test.

These test methods can be used for RAM, ROM, PLA and other logic blocks difficult to test.
However a fault free signature must be evaluated by simulation.

The more of the existing logic that can be utilised for self-testing, the less will become the overhead cost for test.

14.6.1   BILBO

This system employs an internal circuit that can be used as an exhaustive test generator, Scan Path and/or signature analyser.

14.6.2   Linear Feedback Shift Register

Linear Feedback Shift Register - LFSR, can be used to create an exhaustive test pattern.
The pattern will be generated with a shift register where some outputs have been fed back via EXOR gates to the first input of the shift register. When clocked the register will cycle through almost all possible states. This will result in an increase of testability without detailed analysis being necessary to generate a test pattern for all the extra circuitry.

This technique will be frequently used in Signature Analysis.

14.6.3   JTAG Method

Boundary Scan is the test method that connects the internal ASIC chip world to the external printed circuit board or hybrid substrate world for the purpose of monitoring the current logic state of the various connections.
Boundary Scan can be successfully used on larger blocks of logic within the ASIC chip.

As its name implies, Boundary Scan consists of a circuit configuration that, in test mode, allows signals to be passed around the periphery of a system.
The usual implementation will be a shift register that enables the monitoring of the state of many internal connections.
In the test mode, the state on all the connections will be read, then shifted around the system and finally read at an output.

The standard JTAG/IEEE1149-1 is a specification using Boundary Scan Techniques that defines IC and PCB test methods. It is rapidly gaining acceptance within the electronics industry.

It defines a standard 4 wire test bus that can be used primarily to test the interconnections between chips on a PCB.
It specifies a procedure that can be used to test a chip internally.

Test configurations are identified as :-

1. **'JTAG Merchant chip compliant'** if the full JTAG specification is met.
   Standard or Merchant compliant chips can be designed into any PCB application and maintain full JTAG testability ;
2. **'JTAG ASIC chip compliant'** if the chip can be used in an application where all components are under control of the designer.
   Certain reductions can be made in the test circuitry.
   e.g :- Semi-Custom Products.

The minimum requirement to satisfy JTAG will be to include test registers on the input and output paths of a device.
These registers must be connected to a JTAP control cell which uses the 4 wire test bus, as described, to control the registers.
This arrangement does not define an internal chip self-test ability, only how to connect to the board test system through the chip input and output pins.

Testing of the chip must be dealt with by the designer but with certain minimum facilities for external connections being included.

There must be a data path for the signal, 'Test Data In' - TDI, through each chip in order to produce the 'Test Data Out' - TDO.

JTAG merchant chips will only meet the specification if the test registers on the output pins have holding latches on their outputs to prevent upsetting the operation of other chips on the PCB.

14.6.4   Scan Path

The basic principle of Scan Path is to make serial connections between the ASIC inputs and outputs by forming chains of shift registers from the existing circuitry.

The memory elements in the design - flipflops or latches - will be connected in such a manner that, on the command of 'Test Mode', they will be strung together to form a shift register, thus forming the 'Scan Path'.

Forced values of logic states will be clocked along the Path and read at an Output Pin.

Figure 14–1: Overview of JTAG Method

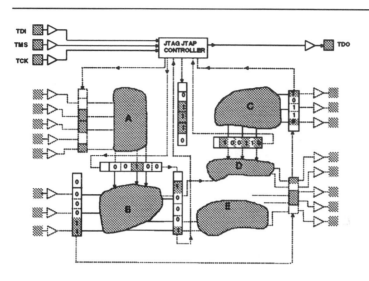

If the pattern emerging is correct, the assumption will be that all
the logic along the the Path is working correctly and, therefore, that
there will be a high probability of the designed circuit performing
correctly when switched to its functional mode.
This will change the testing to appear to be that of a combinatorial
circuit rather than a sequential one and so will reduce the test times.

This may make the use of ATPG more effective.
ATPG is best used with ASICs of high complexity.

Partial Scan can be useful if the area overhead and the performance
degradation using full scan have been found unacceptable.
Partial Scan means that only those nodes that were found
uncontrollable and/or unobservable during fault simulation will be
connected to form a Scan Path.

Advantages associated with full scan test :-

*   It will be easy to initialize the ASIC ;

*   It will be easy to control and observe all internal nodes ;

*   It will be easy to generate testpatterns ;
    ATPG may be used ;

*   It minimises the work for making the ASIC testable ;

*   Only 2 extra pins may be needed.

Disadvantages associated with full scan test :-

*   Extra chip area will be required ;
    5 to 15% depending on the design structure ;
    Due to multiplexed flip-flops and extra interconnections ;
*   The performance of the chip may be degraded ;
    Extra logic and increased interconnections may increase the delay on critical paths ;
*   Testing may not be performed at normal operating speed ;
*   Testpatterns may be longer than the functional ones ;
    the test time increases due to the fact that the tests must be serially shifted in and out of the chip.
    This puts extra demand on the tester ;
*   The test equipment must support Scan Test Methods ;
*   Verification of the signature still remains to be performed ;
*   Testpattern generation and fault simulation may still be necessary on some of the remaining combinatorial logic ;
*   The extra circuitry of the Scan Path may be difficult to test ;
*   A minimum of two extra pins for test will be needed.

14.6.5   Test Synthesis

Test Synthesis divides into two areas :-

*   Testability Synthesis ;
*   Test Synthesis.

14.6.5.1   Testability Synthesis

The function of Testability Synthesis will be to make the ASIC testable. It covers :-

Test Plan Generation

Select a test strategy ;

Plan the ASIC into easy-to-access, hierarchical blocks.

Hardware Insertion

Add extra circuitry to meet test needs ;

Scan or Boundary Paths.

Rule Checks

> Check against Testability Procedures.
> Insist upon the implementation.

Libraries

> Ensure that correct modules and test data are used.

14.6.5.2   Test Synthesis

Test Synthesis will cover :-

Test Data Source Generation

> Generates the bit-oriented testpattern.
> May use ATPG.

Output Formatting

> Translates the testpattern for use by the ATE.

14.6.6   Estimation of Test Time

In production test equipment, it will not normally be recommended that test time exceeds more than two seconds in the test socket.

Test time for a pure digital chip can be calculated as follows :-

$$Tt = Tclk \times Ntvec$$

Tt = Socket test time.
$Tclk$ = Period time of the used clock in the tester.
$Ntvec$ = Number of clock periods in the test input pattern.

In addition, there will be a short time allocated to test the dc parameters of the chip. This will depend upon the ATE used.

14.6.7   Diagnostics

In very complex ASICs, it may be necessary to have testpatterns that not only reject the faulty components in the production test but also have diagnostic capabilities.

If the prototypes fail to meet their specification, the additional circuitry may make it possible to analyse the problems quickly and so suggest remedies. Valuable development time can be saved.

14.6.8   Test Rules

General rules and guidelines for ASIC testing :-

1. Testing must always be considered from the beginning of a design at the highest level of system implementation ;
2. Internal system clocks generated on-chip during normal system operation must be disabled under the test and be replaced by the clock of the tester ;
3. Feedback paths should be avoided in combinatorial logic, if they exist, it will be necessary to break them up during testing ;
4. Redundant logic should be avoided wherever possible and can be reduced by logic minimisation ;
5. Wired logic should be avoided ;
6. Asynchronous designs should be avoided ;
7. Parallel Scan Paths may reduce test times but only at the expense of silicon area ;
8. Very complex ASICs may require the test to be partitioned into several, electrically insulated blocks. Each block will be then tested as a separate, but much smaller ASIC.
   May be adopted with most existing test techniques ;
9. Smalls RAMs may show savings on initial overhead using Scan Test rather than BIST ;
10. Time to silicon may be quicker with structured test techniques ; Economies may be effected later using Ad Hoc testing ;
11. Always design with hierarchical test access in mind ;
12. It is recommended that all inputs and outputs on analogue cells are accessible via pins or multiplexed to one pin ;
13. For some analogue cells it might be necessary to bring out internal nodes to increase the testability ;

14.7   Conclusions

Testing and test methods can often be the most difficult part of the development of any ASIC design.

In very complex circuits, testing requirements may dominate the development and production costs.

Time spent planning the testability of an ASIC will be repaid.

Methods of improving the efficiency of testing and test methodologies MUST be under review continuously.

Test tool assessment must be integrated into the design tools at the highest abstraction level. This will allow 'Fault Grading' to occur at the highest level, more efficiently.

Since testing must be a prime concern at all levels and at all stages of a design, it will become a necessity for the availability and easy usage of test tools and test evaluation methods at all levels.

Not only must test efficiency be improved but also its implementation at the various design stages.

Chapter 15 Design Methodology

15.1 Introduction

A design methodology is not :-

>>> a design tool

> nor

>>> a user manual for a design tool.

It is a recommended order for the use of a specific set of design tools to carry out certain activities.

Due to the diversified nature of the disciplines involved in ASIC design, it is necessary to have a systematic approach to guide the designers through ALL stages involved in development of a system, from the initial idea to the completed product.

It must be remembered that a Design Methodology can not replace design expertise, it can only assist.

Its function is to support the design process and, by a systematic approach, to eliminate many errors and potential errors.

There is not one correct design method.

Any method that can produce correct systems consistently is, by definition, deemed suitable.
Each method will evolve, based on a philosophy of design, tempered by experience and used because it has been found to work for the requirements of a particular System House.

In this chapter, the authors will describe a Design Methodology that has been found to given good results over many designs. It is believed that this methodology is soundly based and will be suitable as the guidelines for a wide range of companies.
Most companies after some experience, will introduce minor changes to match the methodology to their individual design philosophies and/or design styles.

Complex electronic systems have been designed for many years but the introduction of new technologies and design tools have permitted the use of more and more functionality in the systems, mainly through the use of ASICs.
There has been great pressure on the System Houses to adopt the new design tools and technologies as quickly as possible.
A more flexible approach to the introduction of all innovations is essential if a company is to remain competitive.
The ever-increasing rate of change of products has led to their shorter life cycles. In turn this demands shorter product development times. Therefore any ASIC being used in the new system must be correct at the first attempt.

The correctness of the design of all functions must be verified at each step by simulation.

The increased complexity the designer faces can be overwhelming.

It will be necessary for future Design Methodologies on available CAD tools, to support system design based on some variant of

a vertically-integrated, top-down method,

where various stages of the design can be run in parallel.
This methodology must be capable of supporting :-

increased productivity,
multi-level system optimisation
and making the design work more interesting.

This implies, as has been discussed in earlier chapters, the necessity to design, at least initially, in some high-level language.
Although such tools have, allegedly, been 'available' for some years, only comparatively recently have satisfactory ones emerged.
Design tools offering HDL language system descriptions, simulation and logic synthesis have become mature enough to undertake designs other than those of an experimental nature.

VHDL has emerged as the industry standard for a high-level or hardware description language.

As it is now accepted as such, many CAD vendors supply design tools that have the capability for porting designs across systems.
With Silicon Vendors supporting libraries on several design tools, the System House is no longer committed to a single source.

If the new product development timescales are to be achieved, much of what is currently manual work at lower levels of implementation MUST be automated.

High-level modelling, simulation plus synthesis tools for architecture and logic, test pattern generation and automatic layout should speed up the work and also reduce the number of errors introduced at lower levels of implementation during translation.

These techniques will allow time for design exploration and the possibility of improvements in the performance of the system.

As more complicated systems are defined in high-level descriptions and realised in silicon, there will be the added bonus of their reuse in later designs as 'building blocks'.

On more complex ASICs there will be a need to partition the design work between several designers.

The designers will be able to work individually on their own blocks without depending upon the progress of designers of other blocks.

If such partitioning takes place, it is recommended that the specification of each block and its communications to other blocks are verified by simulation before allocating to an engineer. Great care will be needed to avoid introducing system errors at this stage.

The Design Methodology should be capable of alerting the designer as early as possible, to those areas where problems may arise.

The quality of the Design Methodology will be indicated by :-

> an increase in complexity during the design ;
> method is not good enough ;

> Constant or decreasing complexity during the design ;
> method is satisfactory.

The amount of rework that is necessary, due to errors introduced during the design phase, will be reduced to a minimum if the recommendation of verification by simulation is implemented at each level and against the initial system specification where applicable.

15.2   History

Traditionally hardware in systems has been designed by several different teams. One team would be responsible for the total system design. Another team would complete the detailed ASIC design, covering both analogue and digital circuitry and so on. The final team would deal with testing.

The partitioning between the teams helped to reduce the complexity for each designer but it is debatable whether it helped to manage the complexity of the complete system.

One method used to speed up the timescale on complex designs was to put more designers in each team.

However productivity does not continuously increase in a linear relationship with the number of designers in a team.

Figure 15–1:   Productivity Curve

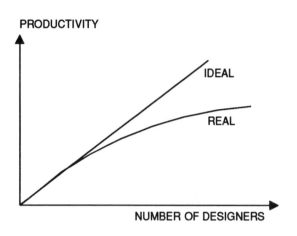

An increase in the team membership from 1 to 2 on one design may result in a large increase in the overall productivity. Two designers make simple operations, like checking, much easier and quicker. They will check each others understanding of the work at each stage.   This will normally lead to a better understanding of the work, result in fewer mistakes and improve the overall productivity. An increase from 2 to 3 or greater, does not appear to give the further, expected improvement. This may be due to management and communication problems within the team. A large team working on one design will probably be at its most efficient if the work is partitioned in such a way as to produce several teams of 2, even outweighing the advantages of parallel design of all modules.

Until recently, few ASIC designers have used Design Methodologies or CAD tools that supported the design flow from the initial specification of the total system to the completed design.
Even fewer have used multi-level optimisation techniques.

Different design activities have been divided into hierarchical areas and the work carried out by separate groups of specialists, see Figure 15–2. Vertical partitioning between system specifiers, system designers and circuit designers may be the cause of misunderstandings arising in the overall project.

Figure 15–2:   Design Group Relationships

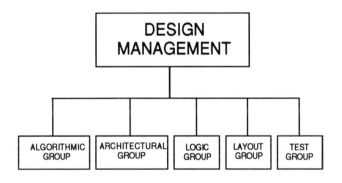

Each specialist group will perform an optimisation within their area of competence and will normally only communicate the result when handing over the information in a written specification.
Time and cost incurred when optimising at one level only, may not improve the overall performance of the system.
In some cases it may even have an adverse effect.

In the type of design system being described, it may not be possible to verify the function of the whole ASIC until it has reached the RTL or Gate level stage.
Whereas most of the decisions that determine the system performance and cost of the ASIC will need to have been taken

Figure 15–3:   Typical Design Flow

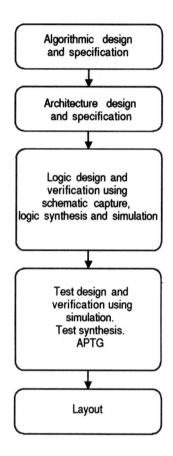

at higher levels, where knowledge of the impact of the changes at the lower levels may be limited.

Detailed verification at lower levels may sometimes be impossible. There may be so much extra data involved in describing the system and storing the simulation results that it precludes simulating the total ASIC on the available computer system.
This will be a risk that will necessarily weigh heavily against the success of the project.
It must also be remembered that the later verification takes place, the more expensive it becomes to correct any errors detected.

Traditional methods have used schematic capture or text netlist description for design entry.
For an ASIC with a complexity of 200000 gates, such a procedure could result in several thousand sheets of description or schematics. The control of that quantity of data would be a major problem.

Many design managers because of their previous experience will still insist upon the use of this type of description - some until recently have carried on with the lengthy and costly construction of breadboards !
Design and productivity progress is thought to be more readily monitored and controlled in these circumstances.

The lack of a stringent design methodology may result in circuit and system experimentation continuing almost until the design needs to be signed off with the Silicon Vendor.

Such methods will result in greater risks being taken as the complexity of the designs rises.

15.3   Vertically Integrated Design Methodology - VIDM

15.3.1   Introduction

The basic ideas behind a new hierarchical high level top-down design methodology will be described in this section.
This methodology should reduce :-

> the amount of design work at lower levels ;
> the amount of debugging ;
> problem solving at lower levels ;

VIDM is :-

> a top-down, vertically-integrated, hierarchical approach.

It should offer the advantage of shorter simulation times.
This will result from operating at higher levels of abstraction during much of the early design period.
It is essential to have continuity of verification through all steps in the design flow.

This methodology has successfully been used by the authors and their colleagues in a number of projects where the targetted requirements were :-

> high complexity, high performance,
> high design productivity and cost effective ASICs.

The method has evolved as a result of necessity, blended with common sense and the experience of many, increasingly more complex designs.

There is nothing magic with this methodology.

Any methodology used in a System House must result in successful developments **IN THAT COMPANY.**

Experience of this approach has show that new designers, after understanding the basic ideas, may rapidly become productive through 'learning by doing' in a design team where at least one member has used the methodology in an earlier, successful project.

15.3.2   Methodology Goals

A successful Design Methodology will require a predefined design flow from the system/specification level to the completed design.

The Design Methodology will have as its major goals :-

* to achieve a first-time correct design ;
* to take the minimum of development time ;
* to incur the minimum of development cost.

For all types and sizes of ASICs, to achieve these objectives, the Design Methodology must :-

* support verification of the design specification by simulation ;
* support functional verification BEFORE detailed design ;
* design a testable ASIC ;
* create confidence in project plans.

In addition on large or high performance ASICs, it must :-

* support design exploration to improve integrity ;
  be feasible to allow several alternative implementations of the system at high levels of abstraction ;
* support multi-level optimisation ;
  starting at highest possible level of abstraction ;
* enable partitioning of work within a design team ;
* make design work interesting and creative.

15.3.3   Methodology Description

The Silicon Vendors offer a wide variation on the level at which they will interface with their customers, i.e. the point in the design flow at which an agreed specification can be signed.

The choice will be dictated by many factors which have been discussed in the chapter 'Interface to the Silicon Vendor'

This choice, together with the operational variations of design tools, has given rise to a number of design methodologies.

VIDM, as stated, is a Top-Down based design procedure.

The example used to describe VIDM is based on the design of an ASIC for a Digital Signal Product.

This design was entered at the System/Specification Level in the form of a DSP algorithm.
Other forms of data and points of entry would be equally valid.

The example will use the four hierarchical levels described in the chapter 'Hierarchical Partitioning into Abstraction Levels'.
The levels were :-

> System/Specification Level, Macro Architecture Level, Micro Architecture Level and Gate Level.

The whole ASIC, including the relevant parts of its environment was modelled to ensure the verification of the ASIC specification.
This should normally reduce interface errors dramatically.

Many test cases were used to simulate the model. Each option was analysed, modified, re-simulated and re-analysed until its overall function was satisfactory.   Identifying and then verifying every possible combination of signals that could appear on the inputs of the ASIC when it is working in its target environment may be impossible due to the complexity.
This was possibly among the weakest parts of the design.
The top model with its environment was then used as a reference against which implementations at lower levels were checked.

The top level model was not finally fixed at this stage.

If any changes were dictated to improve performance at any level, they were implemented and the initial specification updated.
Improvement were defined as a change in one of the operating parameters effecting the performance of the ASIC. e.g. reduction in chip size, power consumed, delay on critical timings or costs.

The whole methodology is based on rapid, inter-level iterations in the pursuit of improvements to the performance of the ASIC.

In this ASIC, the top-down design process resulted in the behaviour of a complex block at the top level being replaced by an architecture(structure) of several blocks with simpler behaviour at the next level down.
The expansion of details from one level to the next ranged:-

between 1 : 10 and 1 : 100.

The actual value depended upon where in the hierarchy the expansion took place. At the higher levels the expansion ratio was small but increased when going down the hierarchy.
More detail on activity at each level is given in the following sections.

The four hierarchical levels as described in this book are capable of managing ASIC complexities up to at least 200000 used gates.

The designers will be responsible for generating the detailed specification of their block, sometimes for the first time. It will however be within the context of the specification for the complete ASIC design.
They will be expected to expand from an initial description, however defined, into a final design that can be satisfactorily simulated in detail.

15.3.3.1   System/Specification Level

A study of the ASIC resulted in a written specification.
The ASIC specification was verified by simulation in its environment, where all relevant interfaces were modelled.
This allowed the system specification to be debugged.

The system was re-simulated with alternative algorithms.
The specification was behaviourally described and verified by simulation, once again with the relevant environment.
The algorithm offering the best option was then selected for implementation. The correctness of the algorithm was evaluated by simulation for as many worstcase conditions as possible.

This strategic part of the design will be vulnerable to error.

Figure 15–4:   VIDM Design Procedure

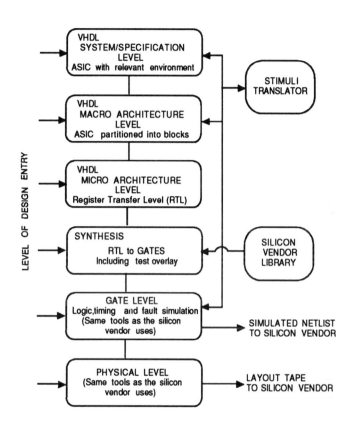

Test was - and always must be - considered at this level.
Timing may be considered but it may be too early for any detailed
analysis at this time.

15.3.3.2   Macro Architecture Level

The specification was found to be too complex for one designer.
It required partitioning into functional blocks of a managable size.
The specification correctness was considered complete after all
partitioned blocks and their environments were verified by simulation
individually and collectively.
Blocks were now designed individually. The work was no longer
dependent upon in-step progress being maintained

Floorplanning was - and must be - considered at this level.

15.3.3.3   Micro Architecture Level

The object was to map the minimum amount of resource to meet
the specified performance.
Resource allocation was performed by mapping the processes onto
architectural resources such as processors, memories and registers.

At this level alternative procedures were applied.
Architectural blocks described behaviourally in VHDL were
converted manually or by synthesis into logic functions that existed
in the chosen Silicon Vendor's macro and cell libraries.

The major step of the implementation was converting RTL to gates.

This was where synthesis was found to make its greatest impact.

Synthesis should be used on all the largest blocks.

Test strategy was selected.

Typically, the trade-offs that had to be considered, were :-

* Clocking strategy ;
* Clock cycle related timing ;
* Selection of parallel or bit serial communication ;
* Selecting optimum word length ;
  for memories, data path and IOs.

15.3.3.4   Gate Level

This is the lowest level of implementation.
Those functions of the system that have not been resolved from a Behavioural Description can be introduced at this level.
This can be directly in gates from the technology library - based on schematic capture, truth tables, etc. The gate level functions were verified by logic and timing simulation.

The design was checked for possible errors, including - race, clock skew, glitches, setup and hold violations.

An acceptable testpattern was generated and verified.

The physical layout was completed using automatic procedures where possible, supplemented by expert manual assistance.

Post layout verification checked that the dynamic performance was still acceptable after layout trackloads were included.

15.4   System Modelling

System modelling for simulation involves, mainly, two major concepts normally referred to as :-

Behaviour and Structure.

Behaviour describes
the functionality of a System or part of the System if it is very large.
It can be seen as an external view of the described object.

Structure describes
how functions in the System or Subsystem are connected.
It can be seen as an internal view of the described object.
The relationship at different levels is shown in Figure 15–5.

15.5   Software and Hardware Partitioning

The VIDM methodology will lend itself to system modelling both in hardware and software.
Any differences in performance that result from different partitioning can be evaluated by simulation and hence used to determine the optimum partitioning between them.

Figure 15–5:   Relationship between Behaviour and Structure

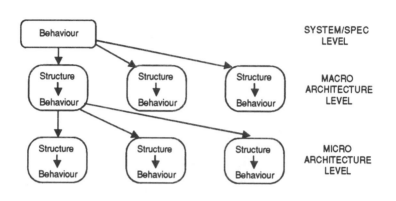

When evaluating the split between hardware and software realisation of functions in a system design, the CASE tools, Computer Aided Software Engineering, may provide a feasible alternative. They may, in some respects, prove easier to manage. Remember

Timing and other system constraints must be controlled in the design sequence before the hardware and software partitioning. If not taken care of at this stage, problems may be encountered later in the design sequence, that are more difficult and very costly.

15.6   Behaviour Coding

The VHDL language has the freedom to describe almost anything that has cause and consequence. It has great flexibility and can support many coding styles. e.g. at one hierarchical level the code for the same function could be written in 10 lines or in 100 lines. The style selected will depend upon many factors and the team must decide the relative importance of the features in their designs. Regularly used, consistent models within a design team will result if recommendations are made on setups and rules for coding in VHDL. A compromise style should be reached from the options :-

> for readability ;
> for accurate and effective simulation ;
> for synthesis.

Remember, at the micro architecture level, it may be possible to synthesise only a subset of the total VHDL language.

The actual subset available may differ between synthesis tools.

The design style, level of entry and exit will have a major impact. Although the VIDM design methodology described, starts at the system/specification level and finishes after the completion of the physical layout, it is possible to enter at any level.

The entry and exit level chosen will depend on many factors.

Some of the more important of which are :-

> Type of circuit ;
> 'glue'* or pure algorithmic ;
> Complexity of the ASIC ;
> If architecture or logic synthesis can be used ;
> Areas of knowledge within the design team.

* 'glue' normally refers to the parts of a system that do not fit into any large functional blocks but are used to 'stick' them together.

15.7   Design Exploration

It can be shown that optimisation will be most effective if it takes place at as high a level as possible.

In most cases, performance and cost optimisations should take place at the algorithmic and highest architectural level.

It is therefore essential that the Design Methodology supports, with a minimum of effort, the generation and evaluation of several alternative algorithmic and/or architectural solutions for the ASIC before the final selection is implemented.

The trade-offs made, should be based on fact, verified by simulation.

If some parts of the design are critical, it may be necessary to implement those totally first, before evaluating other alternatives. Synthesis tools will help to speed up the more detailed implementation of these parts of the system.

'Fixed' functions are those functions specified for the system but defined outside the designer's section of the System House. The current designer will use them but have no control over their operation or design.

Figure 15–6:   VIDM Design Exploration

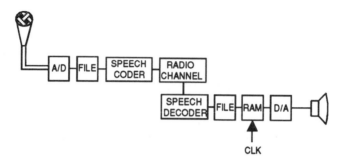

An example of the need for design exploration could be the evaluation and search for an optimum algorithm for speech coding in a radio communication system. See Figure 15–6

Real speech was sampled and digitally stored via a microphone and an analogue-to-digital converter - ADC.

The speech coder, radio transmitter, radio channel, radio receiver and speech decoder could be behaviourally modelled in VHDL.

As the purpose in this case, was to study the speech coder and decoder the modelling effort should be concentrated on these functions. Other blocks could be described in a simpler form.

Typical time needed to write the models for the speech coder and decoder would be in the range of a few man-weeks.

Modifications would take less.

The scenario would be -

> to sample, digitise and store real speech.

The digital speech in this file would then be used as stimuli to the simulator accommodating the models of the different, relevant parts in the system.

After the simulator had processed the speech through the models, the result could be stored in another computer file.

The simulation would not be performed in real time so the result must be clocked out in real time through a digital-to-analogue converter into a loudspeaker.

The designers could by this method, model different coding and decoding algorithms.

The parts would be individually simulated and the results assessed by listening to the quality of the speech.

This would have time and cost advantages over verification with specially built hardware or software programmed DSPs.

15.8   Optimisation

The largest cost reductions or system improvements in ASICs can no longer be achieved by the integration of existing circuitry in existing systems onto one chip.
Nor will it be achieved by trying to minimise chip sizes by 'polygon pushing' in the layout stage.
The Silicon Vendor may be unwilling to consider price reduction.

Any major improvements in solving these system problems will be by the use of design techniques, that employ more effective and radically-new algorithms and architectures.

When total performance requires optimisation and trade-offs cross the boundaries between different design teams and disciplines it is debatable whether the traditional separation of software and hardware design is any longer justifiable.

System optimisation will be a multi-dimensional and multi-level problem. The solutions will very much be a question of how to utilise available resources, - technology, CAD, methodology, human capabilities, time and money.

In multi-level systems, optimisation at the earliest possible stage in the design flow will offer the greatest benefits but does not preclude further optimisation at various levels. This will be true from the point of view of all resources, whether human, money and time and success in this respect will require knowledge of the impact on later stages of any changes made at a high level. This will require designers with tall, thin profiles, discussed in an earlier chapter.

15.8.1   Multi-level versus Single-level Optimisation

Multi-level optimisation has been found to result in cost reductions and performance increases over a wide range.
The values may vary from ten to several hundred percent.
The actual benefit will depend heavily upon the competence of the design team and the structure of the ASIC.

Single-level of optimisation will be normally only based on information concerning the actual abstraction level.

This will be due mainly to the traditional separation of hardware design and optimisation disciplines into algorithmic, architecture, RTL and gate level activities.
In some cases, optimisations are really only partial improvements and may have little or no effect on the overall system performance.

Optimisations attempted at too low a level, may force the design team to consider a solution that involves changing to a more expensive process.

It is, therefore, important that the design team or at least one member, has the knowledge to understand the consequences of any transformations and optimisations being considered between the different hierarchical levels,such as System/Specification, Macro Architecture, Micro Architecture and Gate.

15.8.2   Simulation and Verification

As the number of cells, nets and signals at the gate level increases it may reach the limit where, with the current design tools, it has become impossible to verify the whole system.
It may even become impossible for a designer to get an overview of the total ASIC at this level.
There is a need for a more powerful strategy for verification.
One alternative would be to divide the ASIC into several sub-blocks and then to treat each block as a separate ASIC.
This method has the limitation, that difficulties are experienced in verification of timing on the whole chip.
For optimum performance in system realisation and timescale the simulations should be made at the highest possible level, simulation time will increase approximately by an order of magnitude per step down the hierarchy. See Figure 15–7

The use of a top-down approach and multi-level simulation will enable the designers to verify and correct the design at a higher level of abstraction. System errors should never be left until later as correction at lower levels may prove to be very costly.

Figure 15–7:   Verification Triangle

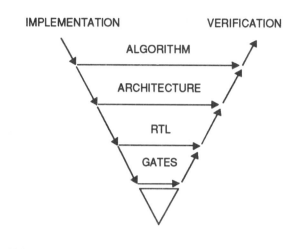

15.8.3   Mixed-level Simulation

Mixed level simulation is essentially offered to enable designers to simulate and verify the function of the whole ASIC with parts described at different abstraction levels. Practically it will mean that those functions that have been implemented already in more detail, can be verified together with other parts still described at a much higher level of abstraction.

The use of mixed-level simulation will mean that after the ASIC has been verified at its top level, it can be resimulated with parts of the design implemented in more detail. At each stage the verification should be against the initial performance at the higher level of representation.
Signal translators may need to be used between these levels if the signal abstractions have been changing.
As the verification strategy can vary from ASIC to ASIC, the designer must fully understand at what hierarchical levels it has been necessary to simulate and with which set of models.
For optimum performance in ASIC realisation and in a given timescale, it will be necessary to make the simulations at the highest possible level for the reasons given in the previous section.

The use of a top-down approach, including mixed level simulation, will enable the designers to verify and correct the design at a higher level of abstraction than has been usual until now.
This methodology will offer the opportunity to verify and evaluate the design behaviour, long before it has been implemented in detail.

15.8.4   Signal Abstractions

An important and difficult area in hierarchical modelling is the concept of modelling signals at different hierarchical levels.
Different levels of signal abstraction will be a necessity if the advantage of mixed level simulation is to be utilised.
The signals at higher levels will need to be abstract types, representations of the contents instead of the physical signals themselves.

These signals typically will be commands and data such as WAIT, ACKNOWLEDGE, BUSY, LOAD, ADD....

Gate level signals will need to work with the physical models.
They will be of logic types such as 0,1,Z,X...

15.9   Top-Down

Top-down design methodology can be used to follow a design flow which starts with the specification of the whole system.

At its highest level of representation, the ASIC to be designed may be represented by one block described by its functional behaviour.

This would enable the simulation of the design in the working environment in which it will finally be expected to operate.

An ASIC design would follow several incremental refinements as more detail was added.
This would continue until the design has reached the appropriate level of representation for the agreed interface to the Silicon Vendor.

15.10   Bottom-Up

Bottom-up design methodology would follow a design flow which started with many simple functional blocks being described.
Normally it would involve the use of schematics, at gate level.

The system design would be based on these schematics.

It would not normally be possible to simulate the overall system, including the ASIC environment.

For large systems this method is considered impractical.

The system description, simulation and analysis would take too long, consume too much computing resource and incur too high a development cost, despite giving accurate results.
It would also increase the risk of system errors.

15.11   Comparison

Neither a pure top-down nor a pure bottom-up design methodology can be totally recommended.

However a top-down approach with additional checking procedures to offset the known limitations at lower levels of implementation, will improve the possibility of generating optimum designs. Especially as more of the routine design work at the lower levels becomes automated.
It becomes a top-down, bottom-up and 'meet in the middle' approach.

15.12   CAD Tools

Design Methodology and CAD Tool flow are not the same.

It is possible to map a design methodology on many different tools.

However it is impossible to completely define a Design Methodology without reference to the CAD Tools on which it is to be mapped.

If the Design Methodology is locked too firmly to one specific tool or tool vendor then the designer may not utilise the best available tools on the market at any one moment in time.

One year, one specific vendor has the best performing tool on the market but next year it may be another supplier.

One CAD vendor rarely has all the tools in a design set that are of equal performance.
If a System House wishes to use the best tool for each operation, there will be a need to 'mix and match'.

Figure 15–8:   Design Flow

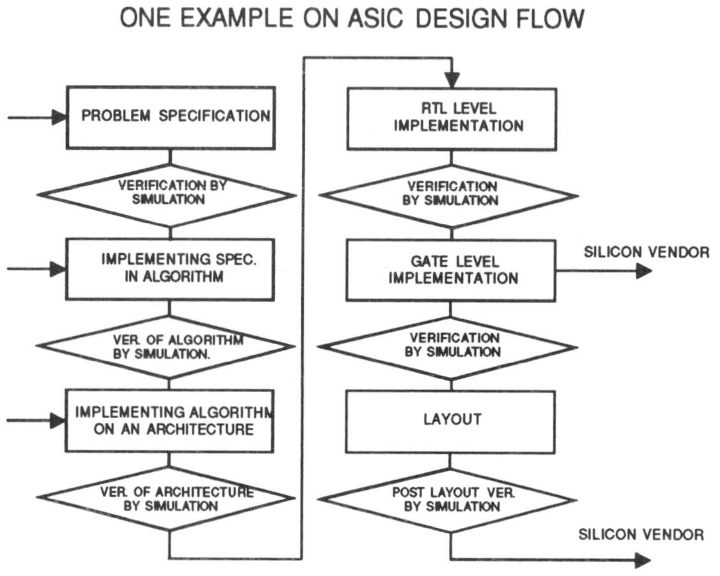

ONE EXAMPLE ON ASIC DESIGN FLOW

```
                                              ┌──────────────┐
                                              ▼              │
  ──►┌──────────────────────┐      ┌──────────────────┐    │
     │ PROBLEM SPECIFICATION │      │    RTL LEVEL     │    │
     └──────────────────────┘      │ IMPLEMENTATION   │    │
              │                     └──────────────────┘    │
         ╱─────────╲                    ╱─────────╲          │
        ╱VERIFICATION╲                 ╱VERIFICATION╲        │
        ╲BY SIMULATION╱                ╲BY SIMULATION╱       │
         ╲─────────╱                    ╲─────────╱          │
              │                              │               │      SILICON VENDOR
  ──►┌──────────────────────┐      ┌──────────────────┐    │      ───────────►
     │ IMPLEMENTING SPEC.    │      │   GATE LEVEL     │    │
     │ IN ALGORITHM          │      │ IMPLEMENTATION   │    │
     └──────────────────────┘      └──────────────────┘    │
              │                              │               │
         ╱─────────╲                    ╱─────────╲          │
        ╱VER. OF ALGORITHM╲             ╱VERIFICATION╲       │
        ╲BY SIMULATION.    ╱            ╲BY SIMULATION╱       │
         ╲─────────╱                    ╲─────────╱          │
              │                              │               │
  ──►┌──────────────────────┐      ┌──────────────────┐    │
     │ IMPLEMENTING ALGORITHM│      │     LAYOUT       │    │
     │ ON AN ARCHITECTURE    │      └──────────────────┘    │
     └──────────────────────┘              │               │
         ╱─────────╲                    ╱─────────╲          │
        ╱VER. OF ARCHITECTURE╲          ╱POST LAYOUT VER.╲   │
        ╲BY SIMULATION        ╱         ╲BY SIMULATION    ╱  │
         ╲─────────╱                    ╲─────────╱          │
              │────────────────────────────│               │
                                              │       SILICON VENDOR
                                              └────────►
```

This is becoming more readily achieveable as international standard interfaces are being agreed.

15.13   Complexity

Normally the management of the complexity in a design will be by a two dimensional separation. See Figure 15–9
The problem will be broken down until it is felt to be manageable. The separation may take place vertically and horizontally.
For example, separate into :-

* Abstraction Levels ;
* Concern.

Partitioning into Abstraction Levels works vertically.

Separation of Concern works horizontally.

Separation of concern means :-

* weakly related functions are separated ;
* strongly related functions are gathered together.

15.14   Software Analogy

High level programming methodologies, using languages like Pascal and C, can offer reduced complexity and increased productivity.

VIDM using a HDL (Hardware Description Language) offers the above improvements.   VHDL is the preferred language, see Introduction to this Chapter. Synthesis from VHDL descriptions, promises ASIC designers a reduction in the mundane task of translating behavioural description into a gate-level netlist by doing most of it automatically, so increasing productivity.

15.14.1   Methodology Audits

It is strongly recommended that all System Houses carry out a Methodology Audit after the completion of each design and even at intermediate stages.

The purpose of an audit will be to check that the Design Methodology has been followed.
It should also review the success of the project and make recommendations with regard to possible changes to the Design Methodology in the light of current and previous experiences.

Design reviews and methodology audits can, with great advantage, be carried out at the same meeting.

15.14.2   Documentation and Specification

One advantage of using the VHDL behavioural description code is that it can be used for the specification at all levels of system description in the hierarchy of the ASIC.
This specification can be used for documentation as well, provided it has been written in a readable form.
The design work becomes self-documented.

**Figure 15–9:   Abstraction   Levels   Versus   Separation   of
Concern**

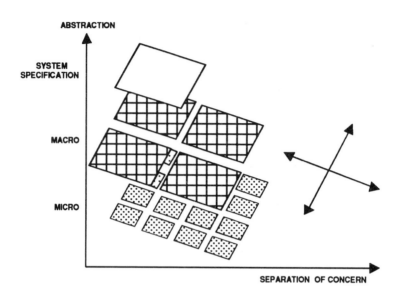

15.14.3   Education and Competence Profile

It is unrealistic to believe that all designers in one team can be
experienced in all disciplines involved.
The more designers there are in a team, the more diverse their
experience is likely to be and hence the more effective their
combined efforts are likely to be.

It will be difficult to work top-down with an inexperienced designer.
This method needs knowledge of the possibilities at lower levels.
It will be a great advantage if at least one designer in the team has
knowledge of all disciplines involved in ASIC design.

The person with the right background will be able to make optimum
trade-offs and will know to whom questions should be addressed, if
problems arise between the hierarchical levels during the design.

Teaching an engineer the skills of using this Design Methodology fully should be a long term investment.
It must be reused by doing subsequent designs where possible.
The management should carefully consider the question of who should be trained, to what level and in which disciplines.
Some designers must be trained as specialists in each of the various design areas described.
A smaller number must be trained on the impact of working in all areas of the overall Design Methodology.

15.15   Conclusions

The Introduction of a Design Methodology will mean spending money and time on training and tools.

The managers must be convinced that the benefits of this method will outweigh the traumas involved in its introduction.

A Design Methodology should support Design Exploration.

Use hierarchy for behaviour, structure and signals.

Use mixed-level simulation to verify that implementations at lower levels still fulfil the initial specification.

Do not introduce any new functionality at lower levels that did not exist at the level above.

An optimum design needs not only good CAD tools and design engineers but also an appropriate Design Methodology.

Designers must give up the attitude that 'it can be fixed later' either with added hardware or the software.

Do not use more hierarchical levels than necessary because they can cause problems when performing multi-level simulations.

The traditional design methods do not work for complex ASICs.

Initial assumptions on methods and tools must be correct in a project - otherwise it can be a painful, costly lesson.

The use of high level design methodology means the decision regarding hardware/software partitioning may be delayed until more information has become available. This should make the decision easier to arrive at and more effective and efficient in its application.

Use multi-level optimisation techniques.

First optimise at the highest possible level where any benefits should be at their greatest.

Use automatic procedures where possible to improve efficiency.

Floorplanning and test must be considered from the beginning.

High level description opens up a whole new world for the re-use of parts or complete designs of functional blocks.

The benefits from learning such a design methodology must be looked upon as a long-term investment

Chapter 16   Physical Design - Layout

16.1   Introduction

In the manufacture of any integrated circuit, the chip will be built up slowly, as the various process stages are completed. The final stages will see the introduction of the interconnection pattern, with as many as four levels of metal being used in some cases.
It will now be possible to see the completed picture of the chip. The diffusions and their relative positions will give the semiconductor devices their particular characteristics and the interconnection pattern will determine the function that is performed.

In design, a reverse procedure will be followed.
In this case, the total picture will be drawn first, it will then be separated into the individual process stages.

Layout is the stage in the design process when the electronic system will be converted into a physical plan of the silicon chip. Until this point in the design, i.e. during the evaluation and simulation stages, the system has been represented only by functional blocks and their interconnections.

In a Gate Array design, only the interconnection pattern will be different, all the other process stages have been pre-determined by the Silicon Vendor.
In other design techniques, e.g. Cell-based or Full Custom Design, will require a complete picture of all the process stages.

Normally the drawing will be produced with the assistance of sophisticated CAD/CAE facilities and as stated above, will depict all the individual process stages brought together. This will produce a complete picture of the chip.

When the drawing has been completed and all the design checks have been successfully carried out, the data will be separated out to give details of each individual processing stage.

The data will be separated to determine the size and relative position of the shapes needed for the various processing stages and interconnection patterns. The shapes will be reproduced, at the photo-engraving stages, on the silicon.

The term Layout or Physical Design will cover

Floorplanning ; Placement ; Routing.

The layout procedure can be carried out

Manually ; Automatically ; or a mixture of both.

The actual format of the layout will vary according to the design style and technology of the ASIC being used.
The two options that will be discussed in this chapter are

Gate Array ; Standard Cell or Cell-Based Design.

In a Gate Array, there is a fixed pattern of transistor groups, peripheral cells and bonding pads.
This fixed part is frequently referred to as the 'Background'.

In Standard Cell there is no restriction with regard to transistor numbers or external connections.   In each case these are determined by the needs of the particular application.

Many of the recommendations will apply to both design techniques.

Any differences will be discussed as they occur.

The layout tools take an input from the designer or layout tools user and detailed design information from the simulation tools database. They produce :-

*   Output for the user during the layout ;
*   Track capacitance data to feed back to the simulation tools ;
*   Physical data used in photoengraving to complete the various manufacturing stages.

16.2   Layout, the Options

The System House may choose to complete a layout themselves or they may choose to subcontract the work.
The options available are :-

Inhouse ; a Design centre ; the Silicon Vendor.

If the decision is take to proceed with the design inhouse, other factors will need to be considered.

The first choice will be dictated by :-

* Time available for product development ;
* Time available to spend on manual optimisation, if necessary ;
* Circuit Size ;
* Density ;
* Critical Path requirements ;
* Specification for power usage.

The second choice will also depend on :-

* Access to layout software ;
* Availability of suitable hardware ;
* Level of experience of personnel to work on the layout ;
* Access to System Designers for consultation.

16.3   Floorplanning

16.3.1   Introduction

Floorplanning is the exercise for determining the possible relationships between all the blocks, cells, I/O cells and bonding pads in a design, whilst considering the electrical and physical constraints that have been imposed on the design.
The main physical constraints will be the design rules.
Conformance with these will be monitored by the design rule checker - DRC - of the particular technology but these will normally be built into the layout software.
The other physical factor in the design of an ASIC, will be to use the silicon area as efficiently as possible, allowing for the disposition of the input and output connections necessary to meet the requirements of the total system.
The Electrical rule checker - ERC - will monitor the correct interconnection of the library functions according to the netlist that was simulated.
Other electrical constraints will include power consumption and propagation delays on critical paths.

In a Gate Array with its fixed background, constraints will also include the total number of available gates and pad positions, together with the likely achievable routing density.

16.3.2   Procedure

Normally the most satisfactory results will be achieved if some floorplanning has taken place before the detailed design began.

A general plan of the chip in outline form of the

        Inputs, Outputs and Major Functional Blocks

should exist as an objective for the design.

The placement of the inputs, outputs, circuitry and available routing channels will effect the success of connecting of power tracks and the flow of signal paths across the chip.
The placement of the blocks effects the success or failure of the routing procedure. The length of the interconnections will also effect the overall dynamic performance of the ASIC.

Some Silicon Vendors offer preliminary floorplanner programs.
These tools are much quicker to use in the initial stages of the floorplanning exercise, because they work with higher levels of behavioural description as input and estimate the size and shape of each function using a statistical method.

Other floorplanners can only work on a netlist input.
They are not usable until much of the detail is known.

Floorplanner programs take the netlist or behavioural information and starting at the bottom of the system, design and estimate the size and shape of all blocks at all hierarchical levels.

This information is then used by the floorplanner to carry out the placement of the system and to estimate the area needed for the interconnections between blocks.

The designer can use this information to more accurately analyse the design before completion of the final layout.

The most effective way and the method likely to achieve the greatest savings is to consider the floorplan BEFORE the ASIC is partitioned. This planning of the general view of the layout of the chip will be a manual operation, carried out 'top-down' at an early stage in the system design before the details are known.
(Automatic floorplanning is essentially a 'bottom-up' procedure.)

Successful floorplanning will result in minimising of the connections between blocks and, therefore, improving the efficacy and efficiency of the routing.

16.4   Layout Modes

16.4.1   Automatic Layout

The designer will input data only to determine the input and output pad positions and, may be, some priority ratings on the order in which the operations should be carried out.

Otherwise the layout tools will be invoked, the system design data presented and the whole operation carried out automatically, with an option for the designer to view the results between operations.

16.4.2   Manual Layout

The human operator will carry out each of the layout stages with an interactive layout editor on a graphical workstation.
Firstly the representations of the cells will be taken from a library and placed in position using the screen. As necessary, they will be moved about by the operator to create the required picture.
The interconnection pattern will then be drawn on the screen, one connection at a time.

16.4.3   Mixed

As its name implies, will use a combination of the automatic tools and manual drawing. The manual operation may take place either before or after the use of the automatic tools at each stage.
If helping before using the automatic tools, the designer will normally place critical functions or 'seed' the placement of some cells to give the autoplacer a lead as to the areas in which particular functions should be placed. Critical connections may also be made.
If helping after using the autoplacer, the manual operation will be to improve the placement in the light of past experience of successful routing with particular layout configurations. This would be before the first attempt with the autorouter.
On other occasions, the automatic tools will fail to complete the entire circuit.
The operator will now have to study the results and attempt to finish the layout by changing the placement, changing the routing or both.

16.4.4   Current Practice

At the present stage in tool evolution, many designs will still require a human help. This will be particularly true when pressing towards the technology limits.

Normally using automatic tools will result in speeding up the layout but this may be at the expense of

> using an increased amount of available area ;
> increasing the delay on critical paths to a totally unacceptable value ;
> increasing the power consumption on the chip.

For manual layout the advantages and disadvantages will be the reverse of those stated above for automatic layout.

Therefore the automatic layout tools have a dual role :-

* to complete a fully-automatic layout, where possible ;
  this will apply to small circuits, circuits without critical paths or those that are not densely packed ;
* to make larger layouts more manageable,
  by reducing the manual operations in non-critical areas.

General points :-

> 1   Inexperienced users should always start with the automatic tools ;
> 2   Automatic tools should be used after synthesis of a large block ;

Synthesised blocks frequently suffer from lack of traceability and automatically generated schematics may be difficult to identify.
This makes circuits that have been automatically generated difficult to layout manually.

These remarks are a reflection of the experience of the authors on currently available software.
The tools on which the experience was based are examples of techniques that are only in the early stages of their evolution.
It is confidently predicted that these generalisations on the tools performance will change.

The introduction of 3 and 4 layer metal interconnection technologies has reduced the need of manual layout dramatically because the success of routing is not as heavily dependent on good placing.
The router has more degrees of freedom in which to operate.

The placement and routing algorithms will improve and, as this happens, they will achieve greater success with the automatic placement and routing of circuits.

16.5   Cells

As an aid to design efficiency, it is recommended that the System House in conjunction with the ASIC Vendor, should keep a library of regularly used, complex, proven macros and system functions.
If the functions are normally critical on dynamic performance, the libraries should include their physical design and routing.

When choosing the level of complexity of the functional blocks used in a design, it must be remembered that using large blocks may speed up the design of the system but make the layout more difficult. Large functions will have large physical sizes and possibly shapes that will be difficult to fit together within the constraints of the ASIC layout parameters.

Large multi-function blocks frequently leave several options unused in any specific application. This may be inefficient in the use of silicon area.

If only small functions are used, although it may be easier to pack more densely, many other advantages will be lost. The design timescale may be considerably increased and the routing pattern of identically functionable blocks may be so different that the dynamic performance no longer matches.

16.6   Hierarchy

Some physical layout programs can not cope with circuits that include hierarchy. All the levels of hierarchy have to be flattened to one level of design.

Automatically placing of identical flattened blocks may cause greatly different placement because of the space available. Therefore the routing patterns may be greatly different.
As a result, the additional loads on the system nets due to the interconnection pattern may be so different that the overall delays through the blocks no longer match.
This may upset critical path performance.

Flattening may also generate large amounts of data which puts a heavy load on the available computer resource.

However, keeping hierarchy unnecessarily at the layout stage may result in the inefficient use of silicon area.

Some layout and routing programs can cope with a design that keeps its hierarchy into the physical design stage.

The use of hierarchy will be frequently needed to manage the complexity of large ASIC designs. The system will need to be described in terms of functional blocks. These blocks will be built up during the system definition and simulation. They will keep their separate identities as the design progresses.
If the blocks are used several times in the total system, each one will need to be given a unique identity.

The critical paths in some large systems may depend upon close matching of the delays through identical functional blocks. If the functional blocks have identical placement and routing, the overall delay through two such blocks will be closely matched.

The use of hierarchy has the advantage of better traceability. Also a repeated functional block will have the same interconnection pattern and matching dynamic performance.
However, even in this type of router system, it may be an advantage, sometimes, to 'flatten' blocks at one or two non-critical, hierarchical levels to improve the packing density and give a better overall chance of a successful layout.

The complexity, i.e. the number of library calls, of functional blocks will vary at different levels in the hierarchy.

Most building blocks will be used at the bottom of the hierarchical tree, when working with gate level functions.

Typically less components will be used to create a functional block when moving up the hierarchy.
The use of behavioural description, coupled with synthesis may cause a variation in this trend.

Some typical figures of block complexities are :-

* Gate level - 100s to several 1000s gates ;
* Top and intermediate levels - 10 to 20 blocks ;
* When synthesised, any level - 100 to several 1000s blocks.

16.7   Placement

Often the success of the routing will be determined during the placement stage, whether this is carried out either automatically or manually.

Automatic placement procedures are now available to take much of the drudgery out of placement.

However when trying to achieve greater than normal packing densities or to improve the delay on critical paths, a limited amount of manual assistance may be necessary for success.

Current practice often includes the 'seeding' of critical blocks.
i.e. a small number of functions may be placed manually before the autoplacer is invoked, to bias results of its placing operation. Connecting points to the blocks may also be chosen manually to ensure that the planned data path pattern is followed.

On a Gate Array, the design of the background, i.e. the fixed pattern of array elements and peripheral cells has been designed by the Silicon Vendor. The number of metal layers of interconnection and the basic cells of a family library will require that a particular philosophy for routing is followed.

For example, with one Vendor the design of cells is such that those to be connected to common clock lines may need placing with horizontal alignment, whereas bus connections between functional blocks require vertical alignment.

In another Vendor these dispositions may be reversed.

The automatic placement algorithms will take these facts into account and so should any manual placement.

In a standard cell design these restrictions will not normally apply.

Automatic Placement is recommended particularly at lower hierarchical levels.   Here the placement of hundreds or even thousands of building blocks may be required.
At upper levels of the hierarchy, the number of blocks requiring placement may be small and can be readily handled with an interactive manual program.

16.8   Routing

16.8.1   Introduction

Routing will cover both signal and power interconnections.
The area of modern chips may be limited by the routing pattern.

Therefore an efficient autorouter and a good metallisation technology are essential features of the offering of any ASIC Silicon Vendor.

The active area, i.e. the area of the active elements used, will often be only a small proportion of the silicon area. For example, an ASIC, with a Sea-of-Gates architecture may, typically, use an active area that is only in the region of 10 % of the total chip area, the rest is routing and unused area.

Power Routing can be a major task in Cell-based or Full Custom design, in Gate Arrays it either exists or is included along predetermined tracks.

Considerable improvements in the amount of noise generated within a system may be achieved by the use of separate power supply systems for the core of the chip and the input and output stages. The outputs may be switching relatively large currents that could upset the functioning of the main system.
As this will need to be planned for early in the design cycle, it is recommended that the topic is discussed with the Silicon Vendor before the detailed design is completed.

Signal Routing is dependent upon the particular design.

The success of a signal routing exercise will be affected by :-

* The capability of the Router Algorithms ;
* The Gate density ;
* The Architecture of the ASIC ;
* The nature of the system design - Random or Structured ;
* The number of levels of interconnection ;
* The placement of the cells ;
* The number of busses and their width ;
* The number of critical paths.

For successful design of any circuit, it will be necessary to use the extracted track capacitances AFTER the actual layout has been completed. These additional loads should be included in the simulation for final analysis of the ASIC before it can signed off for the manufacturing of prototypes.

Any System House planning to complete the layout stage themselves, must assess the capabilities of the placement and routing software they propose to use, as there must be a match between the design approach to hierarchy they follow and that used in the placement and routing software.

16.8.2   Gate Array Architecture and Density

The achievable packing density for a Gate Array will be affected by the factors mentioned in the previous section but principally by the background architecture.

There are three main forms :-

* Channelled Array ;
* Sea-of-Gates ;
* Structured Array.

In the first style, a large percentage of the chip area is plain silicon, i.e. without transistors.
These areas are interleaved between the transistors that are used to form the design functions and the whole area surrounded by peripheral cell components.
The plain areas are allocated for system interconnect.

This means that a high percentage of the transistors can be used but that the total usage of silicon area populated by active components is relatively low.
Frequently as many as 90 % of the transistors will be used in a design but, because of the large areas of plain silicon, the overall usage will only be between 40 and 50 % of the chip area.

In the second style, all the area inside that allocated to the peripheral cells will be covered with transistors, there will be no area left for interconnect alone.

The system interconnection using 1, 2 and 3 separate layers of metallisation, must take place in the areas where the the transistors are not being used to create functions or over the areas where some transistors are being used. In some technologies, interconnection tracks can be run within the areas being used to create functions where all the routing paths are not being used.

This means that a lower utilisation of transistors may take place in general. In some areas however, particularly those occupied by functions with regular structures, very high usage may be achieved. This will result in the the overall improvement in the utilisation of silicon area. Typically 30 - 50 % of the transistors will be used for two levels of interconnection.
If the system, is structured this may increase to between 70-80 %. The utilisation may increase still further as a result of using more levels of interconnection.

The third style will have approximately the same usage as the Sea-of-Gates style in the areas that are of similar architecture.
In the areas dedicated to memory the utilisation will depend on the amount of memory required and how it is organised.

The actual percentage utilisation of any chip will be dependent upon the interconnection pattern required and the number of layers of metallisation available to complete it.

16.8.3   Cell Based Design

This style of design is more flexible than a Gate Array.
No fixed background of transistors exists.
The chip may be laid out according to the restraints of the process design rules, the library components used, the ingenuity of the designer and the effectiveness of the automatic placement and routing algorithms.

16.8.4   Random and Structured Designs

Systems that give rise to structured designs will have regular patterns and, if care has been taken in the design and layout, will normally give compact, successfully routed functions.

Random logic will have little form or pattern and the success or failure of first time routing of any tightly packed areas will be much more problematical.

The allocation of extra area for interconnection between the used cells may be necessary for successful routing but this does not always help and new placements may be required.

Much greater manual intervention may become necessary.

16.8.5   Busses

Busses can consume large parts of the surface area of the chip.

If the bus changes direction, this is particularly true.
Corners - i.e. right angled changes of direction - on a multi-bit
interconnection pattern will require a large area.

It is usually a big advantage to have the functional blocks to be
connected by busses vertically or horizontally aligned.

Wide busses, that is busses with a large number of connections,
that are routed to many of the blocks in a design, will reduce the
utilisation of the silicon area dramatically and should be considered
at the planning stage.
In the case of a Gate Array, this need for extra area may result in the
necessity to go to the next, larger, member of the chip family with its
implication of increased costs. This possibility must be considered
from the beginning of the design and provision made for situation.
There are methods of reducing the number of bits in a bus.
e.g. bit-serial,

     i.e. parallel to serial conversion ;
     distribution of signal round the chip ;
     serial to parallel conversion.

Any improvement in routing requirement by reducing the number
of tracks will need to offset against the use of additional area for
the converters and the increase in the delay on any nets with many
connections.

16.8.6   Critical Paths

The overall performance of an ASIC may depend upon achieving
minimum track lengths when connecting between the particular
functions on the critical path.

The placement and routing may be require careful attention to keep
the additional capacitance loaded on the particular net to a minimum.

Both these operations can be influenced either by manual
intervention or by changing priorities if using an automatic
procedure.

16.8.7   Gate Utilisation

Typical figures for different alternatives within the Gate Array technologies are :-

* 2 layer interconnection - 30 to 70% ;
* 3 layer interconnection - 50 to 90% ;
* 4 layer interconnection - 70 to 100%.

The low figures are fully automatic layout and the high figures with a lot of manual intervention in placement.

16.8.8   Power Consumption

The power consumed on a CMOS chip depends on :-

* number of gates switched ;
* clock frequency of operation ;
* loads connected to the output pads ;
* capacitance of the interconnection pattern.

As the operating clockrate rises, there will be an increasing urgency to reduce the capacitive load by keeping the interconnections as short as possible, particularly on those nets that are very active.

If the chip has a very low power requirement, reducing the load to an absolute minimum will require a substantial amount of manual intervention in the layout stages, once again particularly on very active nets.

16.9   Layout Optimisation

When the designers become more experienced, they may, by the use of an interactive layout editor and manually helping in the placement and routing tasks, gain additional flexibility.
This will give opportunities for greater success in those situations where there is a need for higher utilisation or some other specialised system requirement.

Remember, that system optimisation in most designs is likely to achieve appreciable success only if it has been assessed at as high a level in the work pyramid as possible.

Only in some special instances will it be justified to spend the time improving the layout.

These instances will be to reduce costs and/or to improve the dynamic performance by reducing the connections on critical paths.

i.e. It will be justified if the work does result in a reduction in used chip area. or if the system performance can be improved sufficiently to meet the specification.

If the design has to be realised on a gate-array and the optimisation of the layout results in the circuit fitting on a smaller array in the same family, the production price reduction may be dramatic.

In standard cell, the decision will vary according to the particular project and will be based on

increased development time   v   reduced silicon cost.
   and cost

The reduction in price per unit for a production volume of, say,100k units, may justify the optimisation if the increase in the development cost is tightly controlled.

Silicon prices appear to increase directly with used silicon area over the first part, near the origin, of the price/silicon area curve but tend to increase exponentially after they have reached a critical point.

Optimisation of the physical chip design with its implication of additional manual operations may only be financially viable and give a good return on the effort employed if the chip size lies on the upper, nonlinear section of the price/silicon area curve.
i.e. If the chip size can be brought significantly down the curve, the increased design effort may be suitably rewarded.

In all cases where the layout requires optimisation for whatever reason, it is important that the ASIC designer either carries out the layout or is very heavily involved in the task.

Any layout requiring a high utilisation of available cells or a reduction of the interconnection on a critical path will involve some compromises regarding any proposed changes in the position of various blocks. The designer's knowledge of the system functions will be invaluable in assessing the impact the different options may have on the overall performance of the final chip.

Chapter 17   Mixed Analogue and Digital Designs

17.1   Introduction

As ASIC technology has moved towards higher levels of integration, system designers have found the need to add more and more analogue functions on the same chip as the digital circuitry.
Traditionally when analogue functions have been used in systems, they have been mapped :-

*   with discrete components ;
*   with standard or special chips ;
*   with a few, simple, specially-designed components on the periphery of a digital chip ;
*   designed on ASICs dedicated to analogue functions.

The latter had relatively low component densities.

Where possible, every analogue block was optimised for its particular function and placed close to its interface with the digital circuitry to produce the best-possible overall system performance.
When many analogue functions were integrated on one ASIC, their output positions were dictated by the package. The pad positions that allowed minimum connection lengths were limited. The analogue to digital interfaces were not all optimum. Many of the predicted system advantages of the single analogue chip were minimised by other effects.

If the analogue and digital functions could all be integrated on one ASIC, the system partitioning would be made easier.
A combined function would normally produce a system that was :-

smaller, less power hungry, higher performance,
more reliable and cheaper overall.

The whole system on a chip may be especially advantageous.

There are some major problems to be solved in the area of analogue design, particularly a world shortage of design expertise.
Design tools lag behind those available for digital circuits.
Analogue Test methods and testers require special techniques and are expensive both to develop and to run.
These problems must be solved before the use of analogue and digital functions on the same technology can gain the widely utilised acceptance that is necessary in future electronic systems.

The demand from the System Houses already exceeds the capabilities of the technology and the CAD available today.

ASICs with mixed digital and analogue functions are showing great growth already and a potential that is even greater, once some of the current design and test difficulties are overcome.
Currently the most growth is being forecast to take place in CMOS and BiCMOS technologies.

17.2 Technologies

Processes used for analogue ASICs include

Bipolar, CMOS or BiCMOS.

The latter is a mixture of Bipolar and CMOS.
These technologies are subject to the same development stages as discussed in an earlier chapter. They are graded in a similar way, by minimum feature size - measured in microns or now submicrons.

During the processing various impurities are introduced to create the designed components. Unfortunately they also create some extraneous ones as well. These will include capacitors, resistors, transistors and, to a lesser extent, inductors.
The important and troublesome manifestation in early CMOS technologies was the parasitic 'Silicon Controlled Rectifier' - SCR.
The effect was that under certain circumstances, the chip acted as a relatively low resistance across the power supplies, drawing high current and with the potential of self-destruction.
This was known as the 'Latchup Effect'.
In digital designs, the extraneous effects caused during the manufacture of silicon chips, called 'parasitics', are normally swamped by the signals used, that have step changes.
In analogue designs where signals vary continuously - i.e. by small increments - it may be more necessary to take account of these parasitics because they may have consequences that seriously affect the overall performance.

As these additional features have a lesser impact on the performance of the design than the primary components, they are normally referred to as second and third order effects.

As geometric feature sizes go down in future processes, there may be increasing design difficulties. The parasitic effects in the process characteristics may become more dominant. What were second and third order effects in older technologies may become as important as the first order effects.

At present when mixed analogue and digital systems are to be realised on one chip and there is a reasonable balance between the two parts, Standard Cell or Full Custom design techniques are used, due to the greater flexibility.

This will result in a much higher development charge and longer timescales. All masks must be customised and all stages processed for this chip. As a compromise however, there are some Gate Arrays that have dedicated areas with prefabricated analogue components capable of forming a small number of functions that are limited in performance, complexity and design flexibility.

17.3   Circuit types

Analogue functions available on digital ASICs include :-

op-amps ;	filters ;	band-gap references ;
ADCs ;	oscillators ;	analogue comparators ;
DACs ;	prescalers ;	frequency multipliers ;
VCOs ;	mixers ;	analogue multiplexers;
	modulators ;	phase locked loops

As the digital technology increases its dynamic capability, it is possible that more and more analogue functions will either be realised digitally or included in specially designed areas of the chip. There is a tendency for many existing analogue functions to become simpler and more predictable when designed digitally.

Examples of this trend already include filters and comparators.

The future CMOS Analogue needs may be reduced to a small set of cells, e.g. ADCs, DACs, simpler filters and band-gap references.

17.4   CAD Tools

Analogue CAD tools exist but their ability to handle all disciplines involved in designing analogue cells and ASICs containing such cells in a very complex system is limited or unproven.
Typically the following major tools are required :-

* Schematic capture ;
* Behavioural Simulators ;
* Analogue ASIC simulators ;
  e.g. SPICE and its many derivatives ;
* Mixed-mode simulators for verification of the analogue and digital functions of the whole ASIC ;
* Test vector generation tools ;
* Silicon compilers or module generators for automatic generation of analogue circuitry ;
* Automatic and interactive layout editor ;

17.5   Design Methodology

Traditionally the design of analogue functions in ASICs has been too component-oriented. It always followed a bottom-up approach. A frequent design scenario had analogue functions considered at the highest system level, ignored through the intermediate levels and re-introduced at the transistor level.
Analogue System optimisation was performed at the polygon level. i.e. the sizes and shapes of the diffusion patterns were changed and also their positions relative to one another.
The transistors were designed that met the needs of the current circuit only and could contain transistor shapes with uncharacterised parameters, the performance of which was difficult to predict and was not necessarily any improvement on the system performance.

Analogue designs must cease this precarious approach to design. They must embrace methods similar to those so successfully introduced for digital systems long ago.
The use of proven, library functions should become the norm.
Limiting the use of unproven devices in a design to exceptional circumstances, should improve the designers' efficiency and the predictability of the chip's performance.

Where new devices are included, each step of the implementation must be verified, as accurately as possible, by functional simulation at all stages, starting at the highest level of abstraction.

17.6   Design Flow

A typical design of today follows the flow given below :-

* Schematic capture and circuit design ;
* Generation of functional test vectors ;
* Verification by simulation ;
* Layout ;
* Post layout verification by simulation ;
* Test Program generation.

Schematic capture uses libraries of analogue cells and where necessary basic components.
The rules for placement and routing for analogue circuitry will, in general, be much more critical than that for digital circuitry.
Good correlation between predicted performance and the results from the chip is essential. It will occur only if the models used in the simulation are accurate representations of functions used in the design plus the environment.
It is necessary for the designer to be aware of certain parasitic effects described earlier in the chapter and their practical limits.

17.7   Analogue Design Considerations

During the course of design containing any analogue parts, it is necessary to analyse the impact of parameters that will rarely be considered in a digital design, these include :-

* Element matching ;
* Global process variations ;
* Local process variations ;
* Process gradients ;
* Boundary effects ;
* Noise coupling ;
* Power supply coupling ;
* Substrate noise coupling ;
* Signal noise coupling.

17.7.1   Element Matching

Process variations result in a tolerance in the matching between identically drawn components, placed on one chip.

The matching between identically drawn components will be progressively less as they move from :-

> on the same chip ;
> on the same slice or wafer ;
> on slices within one process batch ;
> on slices from different batches.

The matching achieved in an integrated circuit depends upon several factors and can be good provided the design takes into account the factors that determine the processing variations.
The major factors are discussed below.

17.7.2   Global Process Variations

Photo-engraving

Minor variations in the images created will be due to :-

> differences in the image sizes across the masks ;
> the quality of the linearity of exposing beam ;
> the accuracy with which the images can be reproduced ;
> minor misalignments across the slice.

These effects may be reduced by Stepper Techniques.

Variations in introduced impurities may be due to :-

> uneven distribution of impurity by ion implantation ;
> differences in vertical diffusion ;
> differences in lateral or sideways diffusion.

17.7.3   Local Process Variations

May be due to :-

> Random distribution defects in the base silicon ;
> local presence of dirt on slice or mask ;
> local misalignment ;
> local uneven impurity concentrations.

17.7.4   Process Gradients

As the various processing stages are carried out, there will be minor variations across the slice.
Variations in diffusion depth and oxide thickness may occur.
This will result in changes to the parameters being produced.
e.g. transistor gain, threshold voltage, etc.

17.7.5   Boundary Effects

The actual value of any parameter will be determined by :-

the as-drawn dimensions of the shapes to be produced ;
any edge effects ;
any fringing effects.

The values of capacitance, resistance and the operating parameters of the transistors may all be changed by these effects.

Edge effects do not change linearly with the drawn dimensions.

Process developments that result in the reduction of the dimensions of the minimum feature sizes may increase the percentage change due to the boundary effects.

Matching of components can be improved by the use of optimum layout structures.
Components should be drawn as large as possible so that the variations due to boundary effects are reduced to a minimum.
The components that are required to match, should be physically laid out with the same orientation.

17.7.6   Noise Coupling

Noise coupling may be a major problem in mixed digital and analogue designs. Noise generated in digital areas of the chip may be coupled into analogue areas.
It is not easy to predict the behaviour.
The coupling may take place :-

* down power supply leads ;
* through the substrate ;
* between signal lines.

Power supply coupling

Noise coupling over the power supply is very common. Its effect depends on the interference management techniques employed. These include :-

ground plane shielding ;
separation and multiplication of power supply pins ;
limitation of dynamic load on power supplies
both on and off the chip.

Substrate noise coupling

The substrate or basic silicon slice on which an integrated circuit is manufactured has a relatively high impedance to the ground connection of the system. Noise generated in one section of the substrate may easly be propagated through the substrate to another section of the chip. This noise is capacitively coupled into the analogue components,

e.g. resistors, capacitors and sensitive signal lines.

These effects may be reduced by shielding, using isolating diffusions called - WELLS - under sensitive components.

Signal noise coupling

This is called crosstalk and is mainly capacitive. It can appear when analogue and digital interconnections are routed in parallel. It can be minimised by keeping sensitive analogue interconnections as short as possible and shielded with metal layers like ground. Sensitive analogue interconnections and components should be well-isolated from clock lines.

17.7.7   Good Design Practice

Current practice dictates that in most analogue or mixed designs, much of the design work at each stage is carried out manually or with a great deal of manual intervention.
This is time consuming.
It is error prone and also rapidly becomes boring.

For 'right first time', it is important that the System House acquire data on the behaviour of the analogue cells or, if necessary, to generate the information themselves, in conjunction with the Silicon Vendor.
The circuits can then be verified as functionally correct by simulation.

Where possible, this simulation should be in a real system environment, with representations of any known parasitics included.

The later stages of the design should be considered from the outset. Planning for the various options available for the routing of signals paths and power supply connections must be carried out and their impact on performance assessed.

Mixed digital and analogue designs can be time consuming and on a component-connected basis will normally be more costly than those designs that are purely digital.

17.8   Test

Testing ASICs with mixed digital and analogue functions may be more difficult in production than digital alone.
It will normally require more engineering work and Silicon Vendor interfacing work than pure digital circuitry of a similar complexity.

This is because most complex general purpose test equipment is designed for synchronous digital circuits.

There are no technical reasons that prevents the manufacture of complex mixed-mode testers, there are, however strong financial reasons against the use of general purpose machines of this type.

Most ASIC designs do not have the production volume to justify the capital expense involved in the purchase of a special purpose or dedicated tester.

There are two approaches to test the analogue blocks in a mixed digital analogue ASIC :-

*   Component testing ;
*   Functional testing.

Method One - Component

Each analogue block is tested separately by individual access from the input pins. This may require isolating other functions.
Extra test pins or multiplexing between existing functional pins are normally required.
Testing of each block against component data offers traceability.

This method requires a greater use of silicon area than Method Two. These extra structures may effect the overall dynamic performance.

Can also result in a degree of overtesting and longer test times.

Method Two - Functional

Only tests the function required by the system.

It is performed through the existing signal pins.

Since this method uses no extra test structures, it requires no additional silicon area.

May result in some undertesting due to some loss of controllability and observability.

17.9   Interface to Silicon Vendor

There may be a need for an even closer relationship between the Silicon Vendor and the System House when mixed digital analogue ASICs are to be designed than that for digital ASICs.

The key to the success of the analogue ASIC is not only good, regular liaison with the analogue designers at the Silicon Vendor but also with the process, production and test engineers.
Silicon Vendors undertake continuous process development to improve the chip yield. Any changes may cause parameter shifts that are too small to effect digital circuits but not in analogue circuits. This situation requires continuous monitoring.

17.9.1   Black Box for Total System

A specification for the whole ASIC can be written by the System House and the chip designed by the Silicon Vendor or an independent design centre.

This will normally be the most costly option but, to compensate, as it involves the use of experienced designers only, it may result in less errors and a faster turnround.
The major difficulty will be in achieving a complete, unambiguous function specification.

If the initial specification is correct, the responsibility for the successful completion of the design will not be dependent upon designers at an early stage of the learning curve.

17.9.2   Black Box for Analogue Functions Only

Another approach could be for the System House to give a black box specification of the required analogue cells.   The system specification would be built up from functions, that have been taken from an existing analogue library.

Any functions that do not exist would have to be designed by the Silicon Vendor or independent design centre.

When the design of these new functions, including the physical layout and verification, was completed, they would have to be approved by the customer.

The procedure would then be similar to that established for the use of verified digital cells from standard libraries.

The new functions would be used by the customer to complete the mixed digital analogue ASIC including the final layout.

Normally the test pattern should be developed in a very close relationship with the test engineers at the Silicon Vendor.

Another possibility would be for the System House to give a simulated netlist of the complete system including both analogue and digital functions to the Silicon Vendor or a design centre who would then complete all the remaining work.

17.9.3   Total Design by System House

Design of analogue blocks at the polygon level by the System House would normally be uneconomical because of the detailed process knowledge required. It would also be time consuming.

This route would introduce uncertainty over the development and production timescales due to the learning curve effect.

If this was chosen, it would be necessary to plan with a good safety margin on any project timescale to allow for the possibility of rework.

17.10   Cost performance

Bipolar and CMOS technologies offer advantages and limitations.
A combined version of these technologies called BiCMOS can overcome some of the limitations of the individual technologies.

17.10.1   Bipolar

Some of the advantages claimed for Bipolar Technologies are :-

*   High accuracy in defining parameters ;
*   Good High Frequency performance, (RF) ;
*   High Power drive capabilities ;
*   Low noise.

The disadvantages include :-

*   High Power consumption for functions ;
*   Relatively low levels of Integration.

17.10.2   CMOS

Some advantages of CMOS are :-

*   High level of Integration ;
*   Low power consumption per function ;
*   Lower cost than bipolar.

The disadvantages are :-

*   Less accuracy in parameter definition ;
*   Less drive capability ;
*   More limited analogue capability.

17.10.3   BICMOS

Some advantages of BiCMOS are :-

*   High accuracy in defining parameters ;
*   Good High Frequency performance, (RF) ;
*   High or Low Power drive capabilities ;
*   Low noise.

The disadvantages are :-

*   Lower level of integration than CMOS ;
*   Higher cost than CMOS.

17.11   Conclusions

BEFORE undertaking any mixed analogue and digital design, it is necessary to consider the following :-

1   Is the mixed technology absolutely necessary ?
2   Will the design be well within the process limits ?
3   Is a high degree of element matching possible ?
4   What possible sources of noise exist ?

The main answer to these questions are:-

1   It is almost certainly cheaper to use a single mode technology
2   A design that requires ALL process parameters in one diffusion batch to be favourable, is unlikely to succeed !
3   Use proven designs with matching orientations.
4   Coupling from the substrate ;
     Coupling from the power supply ;
     Coupling from the physical design ;
     reduce signal crosstalk effect by physical separation.

17.12   Future trends

Analogue cell design can, at present, be very slow and costly.
There is a requirement for the generation of analogue cells of many types from high level behavioural description.
This would improve the productivity and simplify analogue design, reducing the error rate dramatically, in much the same way, that simulation and synthesis tools have changed digital design.
This could be changed by the use of analogue synthesis tools.

Later VHDL versions should cover analogue functions and make it easier to deal with mixed analogue and digital circuits at a much higher system level.

If the design productivity is to be improved significantly, better tools and even turn-key design systems for mixed digital and analogue circuits will be needed.

Better design techniques may become more necessary if new processes with even smaller geometries increase the analogue design problems.

As the prediction of the performance of analogue cells is difficult, there will be an increasing need for analogue cells in which the performance can be digitally trimmed.

Some cells of this type already exist to a limited extent.

Examples of digital, field-programmable designs include :-

> switch-capacitance filter ;
> band-gap references.

Better performance analogue cells, that are designed and manufactured to a greater accuracy, are improvements that will be a need and can be confidently predicted as future products.

PART 4

Appendix

Appendix A   Terminology

A.1   Definition of Terminology Used

Each System House will understand its own definitions of the terminology it uses when describing the design of complex electronic systems.

As there may be many variations of these definitions, the authors have given their particular meaning to the words used in this book.

Where the expressions in common usage are unambiguous, they will be used.
In other cases, any uncertainty will be removed by defining the expressions as they have used it.

This glossary defines abbreviations, acronyms, expressions and words as they are used in this book.
They will normally have been defined when they first appear in the text and, in some cases, each time they occur.

Where it is considered necessary alternative meanings of the entries are given in the text.
Where the expansion of an abbreviation or acronym has been deemed self-explanatory, nothing further explanation has been given.
Some terms in the Glossary, may not occur in the book at all but have been included for completeness.

A.2   Glossary

Abstraction
In this context, Abstraction is dealing with an idea that has not been given any detail.
It is a simplified model of a system that concentrates on concepts and relevant characteristics.
Abstractions are used in a hierarchical design methodology. Both circuitry and signals can be abstracted.

Abstraction Levels
The different hierarchical levels at which an electronic system can be modeled with more or less detail.

Active Component
Any component or circuit that introduces gain or direction.
e.g. diode, transistor

ADC
Analogue to Digital Converter.

AF
Audio Frequency.

Algorithm
Mathematical Expression or a sequence of programming steps that defines some function.
May be one of several alternative solutions.
Initially it will not normally include any timing data.

Architecture
A plan of the form and order in which a particular Electronic Function will be realised.   It does not necessarily give detail of the contents.

ASDSP
Application Specific Digital Signal Processor.

ASIC
Application Specific Integrated Circuit.

ATE
Automatic Test Equipment.

ATPG
Automatic Test Pattern Generation.

ATVG   Automatic Test Vector Generation.

Behaviour   A 'Black Box' description of the function or functions to
                    be performed, giving details of the external connections
                    only.
                    Can be functionally expressed in a formal language
                    at a level of description in which the function of
                    the module is expressed using high level primitives,
                    such as arithmetical operators, and not in terms of
                    implementation.
                    It may also include the external timing relationships, from
                    input to output without any description of the internal
                    timing.
                    See also Function.

Behavioural   Any synthesis method from a higher level into a lower
Synthesis   level of Behavioural Description or into gate level.
                    See chapter 'Synthesis' for description

BiCMOS   Combined Bipolar and CMOS technology.

Bipolar   Earliest Integrated Circuit technology. Now tends to be
                    used for special performance ICs. Have advantages in
                    high frequency operation and analogue functions.

BILBO   Built-in Logic Block Observer.

BIST   Built In Self Test.

Bottom-Up   A design that starts in the simplest form using primitive
                    functions. It grows more complex as additional features
                    of the system are included.

Breadboard   Construction of a system or part system from currently
                    available components on a variety of supporting boards.
                    There are several methods used for interconnections.

CAD	Computer Aided Design.
CAE	Computer Aided Engineering.
CASE	Computer Aided Software Engineering.
CD	Concurrent Design.
CE	Concurrent Engineering.
Cell-based Design	Semi-Custom design style.
CFI	Cad Framework Initiative.
Chip	The individual piece of silicon, cut from a slice or wafer, containing the circuit and input/output pads.
Chip Bonding	Connecting chips to a package or substrate, prior to and making ready for wire bonding.
Chip Temperature	Temperature to which the chip will rise under operating conditions. Will depend upon ambient temperature and the rise in temperature on the chip. The rise will depend upon the power dissipated on the chip and the thermal properties of the package, including any heat-sink if used. See also Junction Temperature.
CMOS	Complementary Metal Oxide Silicon.
COT	Customer Owned Tooling.
CPU	Central Processing Unit.
CSIC	Customer Specific Integrated Circuits.

DA   Design Automation

DAC   Digital to Analog Converter.

Design   The process of creating a solution to an engineering
                      specification.

**Design   The recommended order for the use of a specific set of
Methodology**   design tools for a particular technology and design style.

**Design   The specification of the various dimensions of a particular
Rules**   process.   All designs must conform to these rules.
                     Checking for conformance with the rules is normally built
                     into the design software.

DFT   Design for Testability.

Die   See Chip.

**Diffusion   Number of slices processed together, normally exhibiting
Batch**   similar characteristics.

DIL   Dual-in-Line [package].

DRC   Design Rule Checker.

DSP   Digital Signal Processor.

**Dynamic   Active as opposed to passive characteristics.
Performance**   Measure of speed or frequency of an operation.

E-beam   Electron Beam.

EDIF   Electronic Data Interchange Format.

Encapsulate   The process of enclosing a chip in one of many package types.

EPROM   Electrically Programmable Read Only Memory.

ERC   Electrical Rule Checker.

ESD   Electro Static Discharge.

FC   Full Custom.

FPGA   Field Programmable Gate Arrays.

Framework   Concept in which a variety of CAD programs are integrated into an overall system of compatible design tools.

FSM   Finite State Machine.

Function   How an object acts, looked at from outside.
                      See Behaviour.

Functionality   Similar to BEHAVIOUR but is more specific. It describes modules that can be realised in different formats. The final construction will vary according to the specific application. The details will change for low power, high speed, small area etc. The ability to perform one function or several functions combined.

GA   Gate Array - Semi-Custom design style.

Gate   Gate is the most imprecisely used word in ASIC
                            technology.
                            In Gate Arrays, 'gate' frequently refers to :-
                                basic electronic functions - Nand, Nor, Exor etc ;
                                size of a Gate Array - 2040, 100,000, etc ;
                                group of unconnected transistors ;
                                group of transistors connected to form a function ;
                                method of relating system complexity ;
                                5000 Gates equivalent ; meaning the system is
                                equivalent to 5000 2-input Nand Gates.
                            Semiconductor Suppliers give their own names to :-
                            groups of unconnected transistors ;
                            groups of transistors connected to form basic blocks ;
                            groups of transistors connected as larger blocks ;
                            special blocks built from any of the above blocks.
                            See Chapter 'Technology and Techniques'.

Gate   Circuit structure on the inputs of Integrated Circuits,
Protection   designed to prevent destruction of the device by the
                            build-up of electro-static charge.

Hardware   Hardware device which is designed to have its function
Programmable   changed, temporarily or permanently, by the user.

HDL   Hardware Description Language.

Hybrid   Small pieces of silicon or alumina on which can be
Substrate   mounted active and passive components to form a
                            system. The interconnection pattern may use one of
                            several techniques - thick film, thin film, etc.

IC   Integrated Circuit.

IF   Intermediate Frequency.

Implementation   The realisation of a specification in a selected technology.

JTAG   Joint Test Action Group.
                            Standard for test procedures, JTAG/IEEE1191-1.

Junction Temperature
See Chip Temperature.
Usually assumed to be the same as Chip Temperature but may differ due to hot spots.
Process specifications quote a maximum value, above this figure the device may not work or will become unreliable.

KBS
Knowledge Based Systems.

Leaded package
Package with metal pins for external connections.

Leadless package
Package without pins for external connections.
Connections are made to metal printed on the package.

Learning Curve
A plot of the relationship in an operation between efficiency and experience.

Leadframe
Metal part of the package that conducts the connections from the chip through the package to make external connections.

LFSR
Linear Feedback Shift Register.

Logic Synthesis
The automatic creation of a structured netlist with the same functionality as a given Boolean, Equations or Truth table description.

LSI
Large Scale Integration.

LVS
Layout Versus Schematic.

Masks
Glass with chromium picture created by Pattern Generator from a chip design, used in semiconductor manufacture to determine relationship of the various diffusions and interconnections.

MOS	Metal Oxide Silicon, early Integrated Circuit technology - forerunner of CMOS.
MPC	Multi-Project Chip.
MPSD	Multi-Port Scan Design.
MPW	Multi-Project Wafer.
MSI	Medium Scale Integration.
Netlist	Description in a standard format of the interconnection of components to form a system or sub-system, in a manner that is computer readable.
NRE	Non-Recoverable Engineering [cost].
Package	Parts used to encapsulate a chip, to make connection to the chip, to protect it and to enable it to be handled when being mounted on a suitable support medium.
PAL	Programmable Array Logic.
Passive Component	capacitor, inductor, resistor.
PCB	Printed Circuit Board.
PG-tape	Pattern Generation Tape.
PGA	Pin Grid Array [package].
PLA	Programmable Logic Array.

Placer	Program used in the layout stage of a chip design. Automatically puts onto the physical design, the functions that appear in the simulation description of the system being designed.
PLD	Programmable Logic Device.
PLCC	Plastic Leaded Chip Carrier.
QFP	Quad Flat Pack [package].
RAM	Random Access Memory.
Random Logic	A design which has no definable pattern.
RF	Radio Frequency
ROM	Read Only Memory.
Routers	Programs for interconnecting the functions of a system after they have been placed.
RTL	Register Transfer Level.
Scale-of-Integration	Relative measure of number of components on a chip. See SSI, MSI, LSI, VLSI etc.
Scan Path	Test method.
Seal	The final stage in the process of enclosing a chip in a multi-piece package.
Semi-Custom	ASIC design option.

Silicon Foundry	Semiconductor manufacturer that sells surplus processing capacity under contract.
Slice	See Wafer.
SO	Small Outline [package].
Software Programmable	Device which is designed to have its function changed, temporarily or permanently, by a computer program.
SoG	Sea of Gates - Gate Array style.
SSI	Small Scale Integration.
Standard-Cell	Semi-Custom design style.
State-of-the Art	Latest design, technique or technology. Either in late stage of development or early production.
Structure	Describes a level that directly corresponds with the implementation and how the assigned resources are to be connected and how they communicate with each other - i.e. the manner in which the function is to be realised in this specific case.
Structured Logic	A design which has a clearly definable, repeated pattern throughout the circuit.
Surface Mount	The techniques used for assembling packaged chips onto the surface of a supporting medium such as a pcb.
Switch-C Filters	Active filters, the value of which is changed by switching capacitors in and out of the circuit.
System-on-a Chip	A complete electronic system integrated onto one chip.

TAB	Tape Automated Bonding - silicon chip bonding technique.
Test Vectors	Signals that are designed to exercise the function of a particular chip on the ATE.
Top Down	A method used to manage complex designs. The design starts from the highest abstraction level and by successive refinement increases the details as it passes down through the hierarchical levels.
VHDL	VHSIC Hardware Description Language
VHF	Very High Frequency
VHSIC	Very High Speed Integrated Circuit.
VIDM	Vertical Integrated Design Methodology.
VLSI	Very Large Scale Integration.
UHF	Ultra High Frequency
ULSI	Ultra Large Scale Integration.
UNIX	Computer operating system.
Visual Inspection	part of the quality control on processing. After the slices have been sawn into individual chips, microscope inspection takes place to one of several specifications.
Wafer	Basic material on which Integrated Circuits are manufactured. After purification, semiconductors are mixed with N or P type impurity to give the required characteristics. They are grown into ingots and then sliced and polished ready for processing. They are produced in a variety of diameters.

Well	In MOS technology, transistors are formed by the diffusion of one polarity impurity in a substrate of the opposite polarity. In CMOS transistors of both N and P type are created on one substrate. This is achieved creating an effective second substrate by diffusing regions of impurity that are the opposite polarity to that of the substrate. These regions - called 'WELLS' - are used as the sights of the other polarity transistors.
Well	Part of the package in which the inspected chip is bonded, prior to wire bonding.
Wire Bonding	Method of making electrical connections to silicon chips with very fine wire.
WSI	Wafer Scale Integration.
Yield	Number of good chips from 1 slice. Normally expressed as a percentage of good chips to total on a slice.

Index